Managing Wastes from Aluminum Smelter Plants

Managing Wastes from Aluminum Smelter Plants

B. Mazumder
and
B. K. Mishra

WOODHEAD PUBLISHING INDIA PVT LTD

New Delhi • Cambridge • Oxford

Published by Woodhead Publishing India Pvt. Ltd.
Woodhead Publishing India Pvt. Ltd., G-2, Vardaan House, 7/28, Ansari Road
Daryaganj, New Delhi – 110002, India
www.woodheadpublishingindia.com

Woodhead Publishing Limited, Abington Hall, Granta Park, Great Abington
Cambridge CB21 6AH, UK
www.woodheadpublishing.com

First published 2011, Woodhead Publishing India Pvt. Ltd.
© Woodhead Publishing India Pvt. Ltd., 2011

Woodhead Publishing India Pvt. Ltd. ISBN 13: 978-93-80308-13-5
Woodhead Publishing India Pvt. Ltd. EAN: 9789380308135

Woodhead Publishing Ltd. ISBN 13: 978-0-85709-010-2

Typeset by SD Infosystems, New Delhi
Printed and bound by Replika Press Pvt. Ltd.

Contents

Preface

Demand for aluminum metal in domestic as well as in industrial sector has gone up leaps and bounds in last four decades, so is the generation of wastes by these aluminum plants. Aluminum plants generate a number of wastes amongst which red mud and spent pot liners are more prominent in terms of volume of production and their toxicity. Even amongst these last two wastes, spent pot liner is much more toxic than red mud. For these reasons there have been a frantic effort by scientist and engineers in last few decades to develop a process for safe disposal of these toxic wastes. This book depicts all the work done in this area in last three decades and after analyzing these results suggest a method for value addition to these toxic wastes in order to make their decontamination process economically viable.

1

Introduction

1.1   Introduction

Aluminum is an abundant element in earth's crust (occurs to the extent of 7.5% to 8.1%) but being reactive in nature forms oxide and the oxide ore is called "Bauxite". Aluminum metal is being extracted from this Bauxite ore. Table 1.1 lists the basic properties of aluminum metal.

Since aluminum metal is light in weight, reasonably strong (its strength can also be enhanced by alloying with other metals), and having good electrical and thermal conductivity, finds a number of commercial and household applications. Its oxide ore, Bauxite, mentioned above is refractory in nature and thus can not be used directly to extract the metal from its ore. Moreover, since physical beneficiation of the ore does not yield high purity alumina, chemical processing has been the only viable route till today for purification of the raw ore. Bauxite on the other hand occurs in different structural forms depending on the number of water molecules associated with it. For example, while Gibbsite has composition $Al_2O_3.3H_2O$, Bohemite has composition $Al_2O_3.3H_2O$ and Diaspore $Al_2O_3.3H_2O$ (but water insoluble). The chief impurities in all these ores are iron, silicon, and titanium oxides.

Chemical extraction of high-purity alumina from these ores involves grinding the pre-washed ores to a very fine powder (in order to facilitate subsequent rapid dissolution), pressure steam digestion of the powdered mass with caustic solution, filtration (which throws out insoluble materials called RED-MUD), precipitation of pure alumina as $Al_2O_3.3H_2O$ from the clear solution by seed element, and finally calcinations of this precipitate to generate pure fine powder of alumina. Reaction involved in the above steps of purification of alumina can be represented as:

$$DIGESTION:\ Al_2O_3.\ xH_2O + 2\ NaOH = 2\ NaAlO_2 + (x + 1)\ H_2O \quad (1.1)$$

$$PRECIPITATION:\ 2\ NaAlO_2 + 4\ H_2O \rightarrow 2\ NaOH + Al_2O_3.\ 3\ H_2O \quad (1.2)$$

$$CALCINATION:\ Al_2O_3.\ 3\ H_2O \rightarrow Al_2O_3 + 3\ H_2O \quad (1.3)$$

Table 1.1 Physical properties of aluminum metal

Properties	Values
Atomic weight	26.98
Atomic number	13
Density at 25°C	2698 kg/m^3
Melting point	660.2°C
Boiling point	2494°C
Electrical conductance	65% IACS
Electrical resistance 20°C	2.6548 × 10-8 ohm. M
Tensile strength	50 Mpa (7100 psi)
Isotopes (nine numbers)	23–30
Oxidation state	1. (AlH, Al$_2$S, AlF, AlCl, AlBr)
	2. (AlO)
	3. [Al$_4$C$_3$, Al$_2$(C$_2$)$_3$, AlN, AlP, Al$_2$O$_3$]
Electronegativity	1.161 (Pauling Scale)
Atomic radius	118 pm
Electrical resistivity	26.50nΩ.m (20°C)
Thermal conductivity	237 W/mK (300 K)
Thermal expansion	23.1 μm/m K
Young's modulus	70 GPa
Vicker's hardness	167 Mpa
Crystal structure	Face-centered cubic

1.2   Extraction of aluminum metal (Hall-Heroult process)

Since its discovery about 100 years back, the Hall-Heroult electrolytic production of aluminum metal has been the mainstay in aluminum-producing industries all over the world. The process involves dissolving alumina (Al_2O_3) in molten cryolite and electrolyzing the melt using carbon cathode and anode at around 1000–1200°C. Cryolite (Na_3AlF_6) possess the property of dissolving alumina, lowering molten salt bath temperature, as well as possesses good electrical conductance and higher decomposition voltage than alumina. It also has lower density than aluminum metal and thus the extracted metal floats atop the melt during its production. Although cryolite is available in natural forms in some parts of the world (e.g. Greenland), most of the cryolite now used in aluminum industries are of synthetic variety. Synthetic cryolite is

1.1 Aluminum smelter plants worldwide.

being produced by reacting NaF (sodium fluoride) with aluminum trifluoride (AlF_3).

Figure 1.2 below shows the structure of the electrolytic cell used for aluminum production. As can be seen in the figure, the metal cell is lined with refractory bricks inside and then a layer of prefabricated carbon block laid on it which serves as cathode of the electrolytic cell. This carbon layer is called "Pot-Liner". The pot lining carbon slabs are pre-baked from a green body produced by a combination of carbon size +4 to −200 in Tyler mesh in order to achieve maximum packing density in the prepared slab. Raw materials used for fabricating such carbon blocks are basically graphite, anthracite, and binder pitch. Although among these components graphite has the best electrical conductivity, 100% graphite is never used in making the

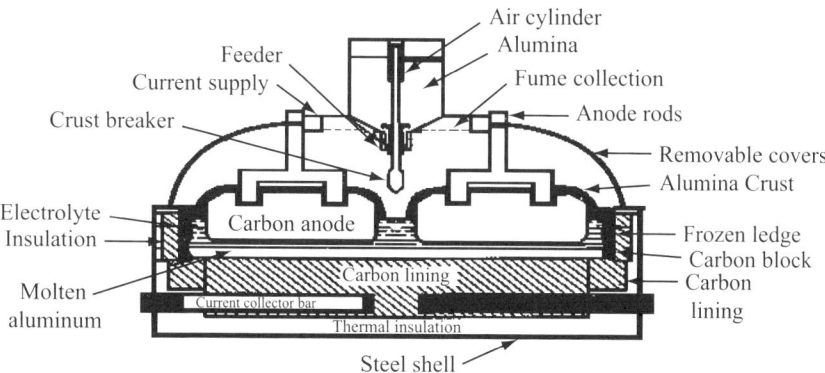

1.2 Schematic diagram of a typical Hall-Héroult cell for electrolytic smelting of aluminum metal [2].

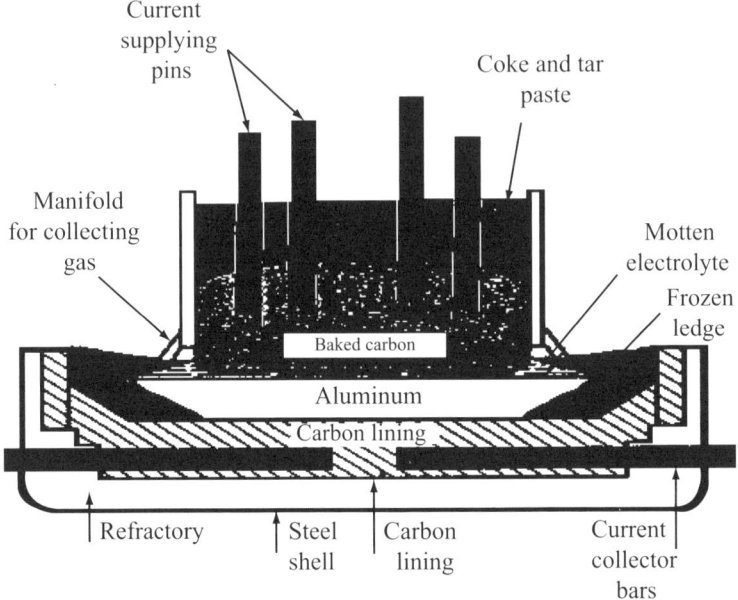

1.3 Schematic diagram illustrating aluminum smelting electrolytic cell with Soderberg anode [3].

cathode blocks as it dose not impart necessary mechanical strength to the cathode block. Various manufacturers use 10–30% graphite in their recipe [1]. Anthracite on the other hand although has short range graphite-like order, its superior mechanical strength imparts necessary mechanical strength to the electrode. Recently some manufacturers are using 100% semi-graphitic carbon to prepare the carbon cathodes. After making the green slab they are baked at desired temperature to produce necessary hardness. Besides preparing such monolithic, carbon ramming masses have also been produced commercially to form in-situ pot linings. They are pressed or rammed on the refractory lining placed on the metal shell of the electrolysis cell, and later burned or heated to achieve necessary mechanical strength.

As regards anodes for the electrolytic cell there are two types of anodes generally used in aluminum smelter plants, and they are pre-baked carbon anode and the other Soderberg anode. Figures 1.2 and 1.3 above show the construction of these anodes. Both these type of anodes are made from the same basic materials and react in the same way. Although Soderberg anode may be more energy efficient, it is easier to treat the volatile wastes if they are not mixed with other emissions (PFC and CO_2) from the pot. Dehydrogenation is often less complete in the Soderberg anode causing more HF to be formed during the anode reaction. Because of this environmental reasons, modern

Table 1.2 Physical properties of cokes used in cathode/anode making

Properties	Values
Density	1.92–2.08 g/cc
Ash	0.15–0.7 wt%
Sulphur	0.1–0.4 wt%
Volatiles	0.2–0.5 wt%
Specific resistance	0.01 ohm/cm

smelters use pre-baked anodes. About 0.5 ton of carbon is used to produce every ton of aluminum. Composition of these carbon anodes are based mainly on petroleum coke and binder pitch. Here again to achieve maximum packing density, a combination of various size fractions of coke is used. Petroleum coke used for this purpose are amenable to needle coke formation in later firing as needle coke, like graphite, offers excellent electrical conductance. The petroleum coke used for this purpose typically have specific resistance of the order of 0.01 ohm/cm, density 1.92–2.08 g/cc, ash content 0.15–0.70 wt%, sulphur 0.1–0.4 wt% and volatiles 0.2–0.5 wt%. After the coke is mixed in specific size fraction with binder pitch (pitch content in Soderberg paste is around 25–35 wt%, and in prebaked anodes 14–17 wt%), it is pressed (5000 lb/sq in.) at 110–160°C to prepare green rods for baking. The green rods are then baked at around 1000–1250°C for 5–8 days and then cooled slowly. During this long heating, the pitch pyrolyses and helps in the formation of needle coke simultaneously binding the petroleum coke particles together. Needle coke having ordered structure, provides superior electrical conductivity. Baked anode prepared by this means is a porous carbon composite with density around 1.45–1.60 g/cc. Typical properties of the binder pitch used for this purpose are shown in Table 1.3 below. Coke structure also plays a vital role in the quality of the anode prepared by this means. More is the crystallized, short

Table 1.3 Physical properties of the binder pitch

Properties	Typical range
Softening point (Ring and Ball method)	100–125°C
Coking value (Konradson)	50–60%
Density	1.18–1.34 g/cc
Benzene insoluble	5–30 wt%
Quinoline insoluble	0–25 wt%
Ash content	0.01–0.3 wt%
Sulphur	0.2–6.5 wt%

range structure of the coke, better is the performance of the finished anode and less is the consumption rate of the anode during metal production.

1.2.1   Other aluminum extraction processes

A number of alternate processes for extraction of aluminum metal from aluminum compounds have been tried in last few decades. However, both technically and economically they could not compete with above Hall-Héroult process, as a result of which Hall-Héroult process has been the only techno-economical viable process for aluminum industries till this date. Nevertheless, it will be useful to mention few of them which have been a close competitor to above Hall-Héroult commercial process for aluminum extraction.

1.2.2   ALCOA aluminum-chloride process

In this process aluminum-chloride dissolved in alkali and alkaline earth chlorides is electrolyzed to obtain aluminum metal. The first commercial plant of 30,000 tones per annum was commissioned and run by ALCOA in Texas, USA in 1976. The starting material alumina for this process is being produced by Bayer Process mentioned above. Alumina is then chlorinated and finally electrolyzed in molten state at around 700–900°C. The inherent problem with this process is that aluminum chloride has a great affinity for moisture which leads to corrosion of the equipments, formation of sludge which affects separation of liberated aluminum metal from the electrolyte phase, sensitive cost of aluminum-chloride production in chlorination step, relative high vapor pressure of several components of the molten mixture, etc.

1.2.3   Carbothermic reduction of aluminous ores

General philosophy applied in carbothermic reduction of aluminous ores involves first step of reduction under conditions where the alumina does not react (the Pederson process) and thus separates it from other subsequent carbon reduction steps [4]:

$$3SiO_2 + 9C = 3SiC + 6CO \text{ (g)} \tag{1.4}$$

$$2Al_2O_3 + 3C = Al_4O_4C + CO \text{ (g)} \tag{1.5}$$

$$Al_4O_4C + 3SiC = (4Al + 3Si) + 4\,CO \text{ (g)} \tag{1.6}$$

The process tends to produce more of (Al + Si) alloys than pure metal (Al content 60 wt%). In subsequent steps vacuum distillation is employed to separate aluminum metal from the mixture. Also extraction of aluminum by sub-halide distillation has been attempted in the past. While the process is technically sound, it throws many economic and engineering (e.g. reactor material etc) challenge toward commercialization of the process.

1.2.4   The aluminum-carbide process

In recent times ALCOA-ELKEM experimented with an advanced reactor process for a larger throughput of metal from a small system there by increasing its techno-economic viability. The process also increases energy efficiency by lowering electric power required to drive the process as compared to contemporary process (e.g. 9500 KWH/ton versus Hall-Héroult 13640 KWH/ton). In this process, aluminum-carbide is being produced from mixture of alumina and carbon, in a specially designed double chambered electric arc furnace. The overall reduction reaction is affected in two stages – in the first stage (at around 1900°C) aluminum carbide is formed as the following reaction:

$$2Al_2O_3 + 9C = Al_4C_3 + 6CO \qquad (1.7)$$

Accordingly, double chamber design and operation brings about better yield (about 67%) by this process. PECHNEY, France conducted design and operation of a double chamber electric arc furnace and terminated the program in 1967 because of higher cost estimate compared to contemporary Hall-Héroult process. However, through improved design in post-2002 period resulted in a cost reduction to the tune of 30% in the reactor alone and further cost-cut through improved energy-saving techniques has resulted in resurgence of interest in aluminum-carbide process. It may be mentioned here that, heating alumina and carbon at atmospheric pressure yield a rather complex product which we can understand from phase diagram of Al_4C_3–Al_2O_3 system shown in Figures 1.4 and 1.5 below.

1.2.5   Reduced temperature aluminum production process

As mentioned above, the present day Bayer process of producing aluminum metal involves a temperature around 1000°C. But recently Argonne National Laboratory, USA improvised a reduced temperature aluminum production process which is currently undergoing pilot plant trial. The ANL process used a modified electrolyte which dissolves alumina and produces aluminum metal at 700°C (about 260°C less than the Bayer process). In its bench scale development process a standard aluminum–bronze anode (non-consumable anode) was used. The anode being non-consumable (unlike Bayer process where a carbon consumable anode is used and produces considerable amount

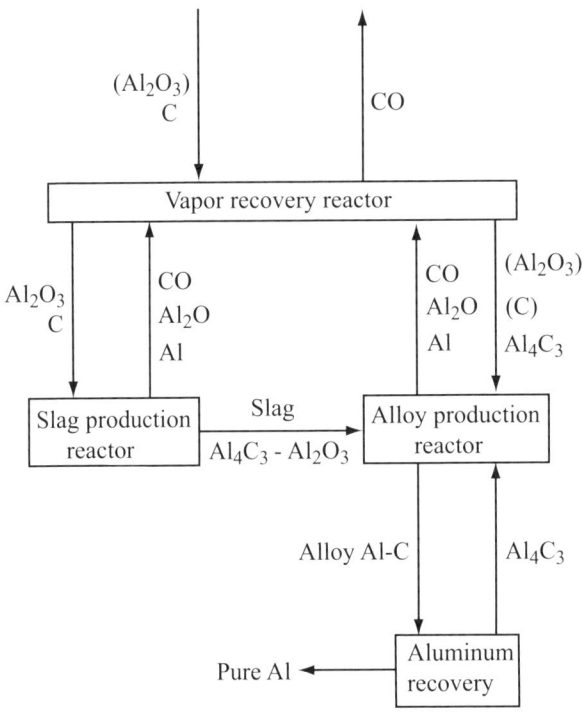

1.4 Flow chart of the ACT-ARP concept for carbothermic aluminum production.

of oxides of carbon) does not release greenhouse gases. This result in higher current efficiency of the cell and no carbon dioxide or perfluorocarbons emitted from the cells. Further, bench scale experiments for more than 100 hours indicate that the electrolyte does not show any significant corrosion at the anode. The electrolytic cells also produce oxygen as a byproduct significantly contributing to the economics of the process. Readers further interested in the process may log on to the website www.tms.org/jom.html.

Permanent metallic anodes

Researches are underway in various research establishments (e.g., central Electrochemical Research Institute, Karaikudi, Tamilnadu, India) for producing non-consumable permanent metallic anodes. These are generally made from refractory metal carbides, nitrides etc. In one such example [5], anode made of cermates having metal content (Ni, Cu, Fe and Cr) to the extent 12–50 volume% have been used to electrolyze aluminum from molten fluoride salt. The ceramics are generally ferrites and very much corrosion resistant in

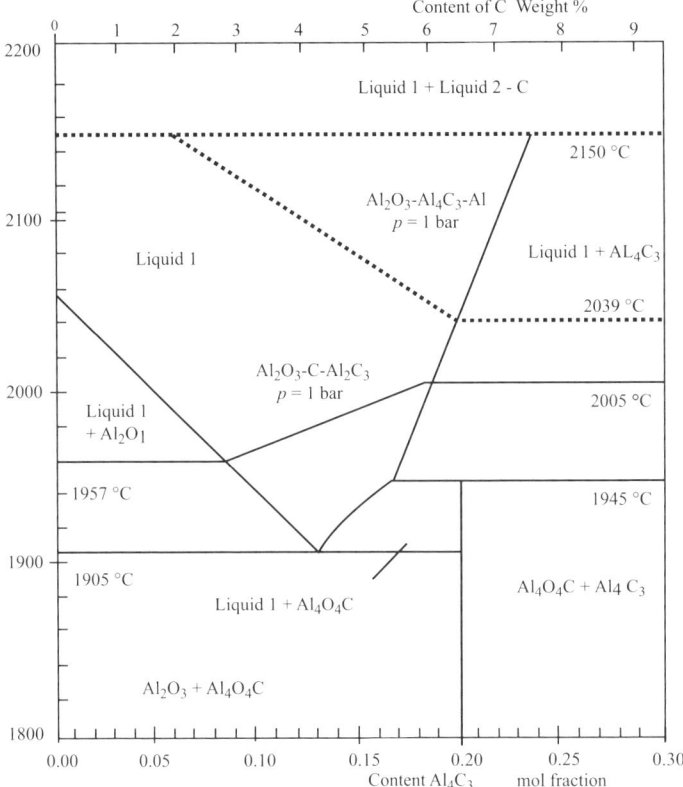

1.5 The pseudobinary phase diagram for Al_4C_3–Al_2O_3 with equilibrium temperatures for the main phase combinations [7].

fluoride environment. In yet another attempt [6], an oxygen evolving anode has been tried to electrolyze aluminum from molten cryolite. The anode comprises of a container made from an alloy of aluminum and at least one metal more noble than aluminum; a fluid bath in the bottom of the container having the ability to dissolve aluminum, said fluid having a density that is higher than the density of molten aluminum at the operating temperature of the cell; a pool of molten aluminum floating on top of the fluid bath in the bottom of the container; a refractory layer arranged on the inner sidewalls of the container in the area of the pool of molten aluminum. The refractory layer protects the molten aluminum from contacting the inner sidewalls of the container.

1.3   Wire-rod drawing plant

One of the ancillary units commonly commissioned within the premises of aluminum smelter plant is aluminum wire-rod drawing plant. Freshly prepared

1.4 General composition of waste emulsion from wire rod mills of aluminum plants

Oil content	Pesticide/Fungicide	Aluminum metal powder
10–12%	4–5%	1–2%

aluminum cast slabs are drawn into 1–3 mm wire and then bundled together to form a composite electrical wire for commercial use. Both, because of good electrical conductivity and lower price compared to conventional copper wire, aluminum wires are finding increasing applications in both house wiring as well as in long distance electrical transmission. The wires are drawn through a hard die which is being continuously lubricated by special grade emulsion oil. NALCO, India uses oil-in water emulsion generated at the plant site by mixing AROL grade lubricating oil produced commercially by M/s Indian Oil Corporation, with water and some additives. For long storage life some insecticide/fungicide is also being added to the prepared emulsion. These emulsions are highly stable and used in the wire-rod drawing machine by pumping them up the rolling stand. The run down emulsion are continuously collected, filtered (to remove aluminum powders), and pumped back again for reuse. These emulsions at the end of their service life are discarded and the discarded waste emulsion contains 10–12% oil in water. Since they also contain sulphur containing insecticide/pesticide, their disposal is a challenge from environmental consideration. Thus it is necessary to break the waste emulsion and bring down its oil content to as low as 10 ppm or lower, for final discharge of the residual water to the open drain. We will discuss methods of cleaning these waste emulsions for release of clean water in later chapter.

Wastes generated by aluminum smelter plants

2.1   Introduction

The main wastes generated by aluminum smelter plants are as follows:

(a) Spent pot liners
(b) Anode butts
(c) Red mud
(d) Coal powder/rejects
(e) Waste emulsions
(f) Coal ash (both fly ash and bottom ash).

Some of these wastes are enormous in quantity simultaneously toxic in nature. For example, for every 1000 ton of aluminum production, 20 ton of spent pot liner is being generated. Similarly red-mud is also a voluminous waste material generated by the alumina production plants. Captive power plant of NALCO (India) generates about 4400 ton/day ash (input coal of 10000 ton/day). In this chapter we will discuss nature of these wastes and their formation mechanism as well as ways and means to clean them and possibilities for value addition.

2.2   Spent pot liners

Average pot life of a aluminum smelter plant is 3–10 years. During their service life, the pot liners (which are heavy carbon cathode blocks laid inside the electrolysis cell) remain in direct contact with the two most corrosive melts at around 1000°C. These are cryolite and liquid aluminum metal. At high temperature aluminum reacts with carbon forming aluminum carbide:

$$4\,Al + 3\,C = Al_4C_3 \qquad (2.1)$$

Formation of aluminum carbide at the contact surface prevents aluminum–carbon reaction, but if cryolite is present, the insulating layer gets dissolved and the reaction becomes continuous. Further, if alumina concentration is above 4 wt%, alumina oxycarbide is formed:

$$Al_2O_3 + Al_4C_3 = 3Al_2OC \text{ (s)} \tag{2.2}$$

Aluminum-carbide also gets dissolved in the molten electrolyte as well as in the liberated aluminum metal. Maximum amount of aluminum carbide is being formed in the alumina-free electrolyte. Thus, near the side wall of the cells concentration of alumina tending to be low, formation of aluminum-carbide is favored. On the other hand, the bottom sludge being rich in alumina, carbide formation at the bottom surface is enhanced. The sludges are created due to excess addition of alumina. These sludges contain about 40 wt% alumina (corresponding to 20 volume% solid alumina) and exist as heterogeneous solution of alumina powder with molten cryolite. Alumina is generally added as a regular shop floor practice for breaking crust formation and is being removed batch wise at every 2–4 hour interval. While only 1 wt% of alumina goes for breaking the crust immediately, due to restricted supply of heat content of the bath available in short time, rest of the unreacted material deposit as sludge at the bottom of the cell. If the cryolite is stirred during alumina addition, the dissolution time has been found to be zero order of kinetics with respect to alumina content of the melt up to 5–8 wt% alumina at 1020°C and with 4–5 wt% alumina content dissolution found to be 0.07 g alumina per minute. Also chemical analysis of failed cathode blocks (spent pot liner) indicates that there is variation in chemical composition of the molten electrolyte and consequent variation in composition of penetrating electrolyte into pores of the carbon block. Typically, the later has well developed crystals of sodium fluoride (NaF) in the adjoining veins of aluminum carbide. Thus the change can be depicted by following equation:

$$4\, Na_3AlF_6 + 12\, Na + 3C = Al_4C_3 + 24\, NaF \tag{2.3}$$

Further, oxygen from air helps build up crystals of alumina (Al_2O_3) as per following reaction:

$$4\, Na_3AlF_6 + 12\, Na + 3O_2 = 2\, Al_2O_3 + 24\, NaF \tag{2.4}$$

Leachable fluoride content in SPL varies upto 20%. Besides sodium fluoride other fluorides found in SPL are Na_3AlF_6 and CaF_2.

The next most important reaction which leads to the failure of pot liner is the sodium-carbon reaction. Sodium tends to enter the cathode lining through formation of intercalated compound. More specifically, the main sodium uptake has been found with the binder coke matrix. Here the reaction involves formation of lamellar interaction compound like C_8Na and $C_{64}Na$. Formation of these intercalation compounds causes increase in molar volume and consequent increased internal pressure within interstices, helping propagation of cracks within the pot liner. However reaction of sodium with graphite carbon is much more difficult than that with amorphous anthracite or petroleum coke. It has been established by experiments that during operation of the electrolytic cell, progressive graphitization occurs in the carbon lining, although recrystalization is catalyzed in the presence of metallic sodium.

Sodium also helps in the formation of sodium-cyanide with atmospheric nitrogen catalyzed by the presence of iron compounds:

$$2\,Na + 2C + N_2 = 2NaCN\,(I) \hspace{2cm} (2.5)$$

Another compound found in the spent pot liners but in minute quantities is aluminum- nitride. This compound could also contribute to formation of sodium-cyanide, especially when there is a continuing source of sodium and air. Other cyanides found in SPL are $Na_4Fe(CN)_6$ and $Na_3Fe(CN)_6$. Besides above contaminants found in SPL in smaller quantities are arsenic and polyaromatic hydrocarbons.

A typical analysis of spent liner collected from NALCO smelter plant, Anugul (Orissa), shown in Table 2.2 below:

Table 2.1 Chemical composition (%) of fly ash and spent pot lining [8]

Serial No.	Name of the plant	Spent pot lining	Fly ash
1	HINDALCO, Renukoot	Al_2O_3: 19.0–24.0; SiO_2: 10.0–13.0; Fe_2O_3: 0.5–0.8; Na_2O: 27.0–34.0; CaF_2:2.0–4.0; fluoride (T): 11.0–14.0	Al_2O_3: 13.0–35.0; SiO_2: 53.0–71.0; Fe_2O_3: 3.5–12.0; Na_2O/K_2O: 0.12–5.0; P_2O_5: 0.3; MgO: 0.28–3.24; SO_3: 0.005–1.1
2	NALCO	SPL: middle layer; Al_2O_3:1.88; SiO_2: 1.56; Fe_2O_3: 0.53; Na_2O: 19.22; fluoride: 10.88; CN: traces 0.3; LOI (C): 67.61 SPL: bottom layer; Al_2O_3: 3.2; SiO_2: 65.81; Fe_2O_3: 1.18; Na_2O: 1.24; fluoride: 2.5; CN: traces 0.08; LOI (C): 0.54	Al_2O_3: 28.23; SiO_2: 59.72; Fe_2O_3: 5.37; CaO: 1.39; Na_2O: 1.23; P_2O_5: 0.93; MgO: 0.83; SO_3: 0.21; MnO: 0.08
3	General analysis for other Indian aluminum units	Al_2O_3: 15–25; SiO_2: 9–13; Fe_2O_3: 0.4–0.9; fluoride: 12–20, CN: traces 0.08; LOI (C): 40–55; CaF_2: 0.123–1.23	

Table 2.2 Typical contaminants in spent pot liners

Contaminants	Typical range (wt%)
Alkali	11–20
Aluminum	5–10
Fluoride	5–20
Cyanide	0.12–0.20

As can be seen from above table, major toxic contaminants in SPL are fluoride and cyanide. Determination of both fluoride and cyanide requires specialized techniques and apparatus estimation of these two anions. Quantitative estimation techniques for fluoride and cyanide determination are enumerated below:

(a) Fluoride

First of all, it is important to recognize that fluoride occurs as both leachable and non-leachable form in spent pot liner. It is the leachable form of fluoride which gets washed out by water and contaminates potable ground water. Total fluoride reported in the literature for spent pot liner is the sum total of leachable and non-leachable fluorides. Since fluoride determination depends on solubilization of fluoride from spent pot liner to filtrate, it is imperative that non-leachable fluoride needs to be converted to soluble form for estimation of total fluoride in spent pot liner.

SM 4500-F/SW-846 method 9056 (ASTM) can be used for determination of total fluoride in spent pot liner. Non-leachable fluoride is extracted into water phase either by alkaline fusion (Sodium carbonate, Zinc oxide) [9] or by fusion with calcium hydroxide and sodium hydroxide mixture [10, 11]. Recently a borate fusion method [12] for determination of fluoride in coal has reported by Wood et al. This procedure can also be applied for determination of fluoride in spent pot liner. The procedure incorporates in two stages sinter-fusion procedure using lithium carbonate, lithium tetra-borate and zinc oxide. In the sinter stage, the carbon mass is ashed at 600°C and the lithium carbonate as well as zinc oxide act as collectors to capture volatile species of fluorine. In the second stage, the ash residue is decomposed by fusion with lithium carbonate and lithium tetra borate at 1000°C. The fused material is dissolved in nitric acid and the fluoride in solution is determined by using an ion selective electrode. Method ASTM D3761-

Table 2.3 Typical fluoride content in SPL samples produced by American Smelters [10]

SPL sample from	Leachable fluoride (mg/l)	Total fluoride (mg/l)
Alcoa, NY	3,210	66,400
Alumax/East Alcoa, MD	3,000	69,600
Naranda Aluminum, MO	3,070	57,700
Ormet corporation, OH	3,230	57,400

96 [13] also describes a standard test method for determination of total fluoride in coal by oxygen bomb combustion/ion selective electrode method. However, lower limit for fluorine determination is only 20 mg/g sample.

Leachable fluoride can be determined by ASTM method D3987-85, where a representative aliquot of 70 g of solid sample was placed in a 2000-ml container, added 1400 ml of water at room temperature and stirred the suspension for 18 hours. After filtration the extract is submitted to distillation and fluoride evaluation carried out through the ion selective electrode method.

Table 2.3 below indicates typical values of fluoride analysis for four spent pot liner samples obtained from four American smelter plants.

Present authors have measured leachable fluoride coming out of NALCO smelter plant (Orissa, India) [15] and contaminating surrounding areas. The result indicate that in most of the adjoining villages of smelter plant, fluoride containing in potable water are within WHO permissible limit, but in few villages fluoride contents are above permissible limit. Similar study has been conducted around the BALCO smelter plant (Korba, India) by Sahu et al. [16] and similar conclusion was drawn by above authors. It may be noted in above table that total fluoride has been expressed in w/w ratio (mg fluoride/kg SPL).

(b)   Cyanide

Cyanide in spent pot liner are reported both as leachable and total cyanide. Total cyanide extraction is performed according to the procedure established by EPA SW 846 method 9.010 B [17] with a solid sample size of 10 g and a distillation time of 1.15 minutes. Leachable cyanides are determined using EPA SW 846 method 9.013 [18] where a representative aliquot of 25 g solid sample is placed in a 1 liter bottle to which 500 ml of water and 5 ml of 50% (w/v) of sodium hydroxide aqueous solution are added. The mixtures stirred at room temperature for 16 hours. The pH of the extract is maintained above 12 throughout the extraction step and subsequent filtration. After filtration the extract is distilled as per procedure established by EPA SW 846 method 9.010B and the distilled cyanides are evaluated using an ion selective electrode.

2.3   Anode butts

Anode butts are the residual electrode left in the electrode holder at the end of electrolysis with prebaked carbon anodes. Anode consumption rate in aluminum cells are 334 kg carbon/ ton aluminum produced (100% current efficiency) and in regular practice around 0.41–0.48 carbon/kg metal is consumed. The anode carbon is being consumed by ohmic resistive heat and electrolytic oxidation. However the part of anode just above the melt is being continuously subjected to high temperature reactions with fluoride melt and atmospheric oxidation, which ultimately results in impurities left in the discarded anode butts. Table 2.4 below shows typical contamination range in such anode butts.

At the end of their use, the left over part of the anode (called "butts") are crushed and recycled to produce new anodes. However, while recycling the butts the built-up of sodium concentration is closely checked and controlled in order to avoid harmful effect of excess sodium in newly formed anodes.

Table 2.4 Typical contaminants in anode-butts

Contaminants	Range
Sodium	0.05–0.65% (average)
Fluoride	2.0–3.5%

Table 2.5 Typical properties of commercially available prebaked anodes

Property	Value
Ash	0.5–10%
Specific resistance ($\mu\Omega$m)	55–60
Compressive strength(M P_a)	30–45
Bulk density	1.5 g/cc
Sodium content	100–300 ppm
Coefficient of thermal expansion	0.4%
Flexural strength	9–12 MP$_a$
CO_2 reactivity	dust% 2.5
	residue% 88
Air reactivity	dust% 92
	residue% 2
Thermal conductivity	3 W/m K

Good quality anodes should be oxidation resistant in gaseous environment in order to minimize oxidation on the expose surface ($C + O_2 \rightarrow 2CO$) [19]. It has been found that [20] an average increase of sodium content by 48 ppm causes average increase of 3.38% and 2.72% increase in air and CO_2 reactivity, respectively. It has further been reported that sodium content increases from 127 to 367 ppm with addition of butt powder from 5 to 25%. Commercially available prebaked anodes have sodium content ranging from 100 to 300 ppm [21]. Anode butts also contains considerable amount of fluoride. Present authors checked fluoride and sodium content of anode butts collected from one of the largest aluminum smelters plant in India (NALCO Ltd) and treated them by special chemicals. Results are shown in following Table 2.6.

Anodes are made from CPC (70–80%) and binder pitch (30–20%). Typical properties of CPC and binder pitch are given in Table 2.7 and 2.8 below:

Table 2.6 Sodium and fluoride content of anode butts obtained from NALCO smelter plant (India) and drop in sodium and fluoride content by treatment with special acid mixture by the authors [22]

Sample	% F⁻ content in anode butts	% Na content in anode butts	% F⁻ content after treatment	% Na content after treatment
1	3%	0.55%	0.8%	< 0.003%
2	2.5%	0.60%	0.79%	< 0.003%

Table 2.7 Typical properties of coke used in anode making

Property	Raw coke	Calcined coke
Specific resistivity (Ω cm)		
Ash (wt%)	–	0.009–0.011
Volatile matter (wt%)	0.1–0.7	0.15–0.70
	5–15	0.2–0.5
Bulk density (g/cc)	0.6–1.1	0.65–1.12

Table 2.8 Typical property of pitch used as binder in anode making

Property	Value
Softening range	85–110°C
Cracking value	56

Continued

Ash content	<0.3%
Density	1.59 g/cc
% Insoluble in toluene	15–30%
% Quinoline insoluble	< 10–8%
Beta resin	>18
Electrical resistivity	66 $\mu\Omega$m
Thermal conductivity	6 W/M °K

As mentioned earlier in the recipe of anode preparation, spent anode butts are mixed keeping in view required range of properties for prebaked anodes, so that the anode works satisfactorily in service. This in turn implies restrictive addition of recyclable anode butts to the fresh batch of raw materials for production of prebaked anodes. For this reason most of the smelters have restricted use of butts maximum to the extent of 5% in regular recipe.

Generation of anode-butts by aluminum smelters are quite high (about 0.5 ton anodes are consumed for every ton of metal production) and difficult to dispose in open field for possible environmental pollution. Recirculation of the butts involve crushing, washing and classifying into various size fractions in order to maintain size distribution demanded in the process of anode manufacture. Most of the contaminants in the anode-butts occur on its surface, although some fluoride and alkali may penetrate into its body due to generation of porous structure during high temperature electrolysis. While the binder pitch used in the manufacture of anodes are predominantly anisotropic and thus gets graphitized by pyrolysis at elevated temperature of the bath, its amorphous carbon content however revolves around 0.07–4.5 wt%.

2.4   Red mud

As mentioned earlier (in the conventional Hall-Heroult process of aluminum extraction) red-mud is produced in Bayer method of extracting alumina from aluminum ore and the amount of red-mud produced is almost same as alumina produced. After extraction of alumina by alkali leaching, the muddy gangue material left is called red-mud.

Although its composition varies as per the source of Bauxite used [23], its main constituents are generally ferric-oxide, silica, alumina (untreated), titania, and the residual alkali. Table 2.9 below indicates the range of these constituents generally found in the red-mud produced in various aluminum production plants round the world.

As the above table shows, red-mud contains large amount of alkali and attempts have been made to reduce this alkali content either by use of microbes [24] or by chemical means [25, 26]. Waste steam from thermal power plant

Table2.9 Red mud constituent from various aluminum smelter plants in India

Company	Constituents				
	Fe_2O_3	Al_2O_3	TiO_2	SiO_2	Na_2O
Al. Corpn.	20.26	19.60	28.00	6.74	8.09
MALCO	45.17	27.00	5.12	5.70	3.64
HINDALCO	35.046	23.00	17.20	5.00	4.85
BALCO	33.80	15.58	22.50	6.84	5.20
NALCO	52.39	14.73	3.30	8.44	4.00
Hungary	38.45	15.20	4.60	10.15	8.12
Jamaica	50.9	14.20	6.87	3.40	3.18
Surinam	24.81	19.00	12.15	11.90	9.29
ALCOA mobile	30.40	16.20	10.11	11.14	2.00
Arkansas	55.6	12.15	4.5	1.5–5.0	Traces
Sherwon	50.54	11.13	Traces	2.56	9.00
FRG Baudart	38.75	20.00	5.5	13.00	8.16

Table 2.10 Typical ranges of constituents in red-mud

Constituents	Range
Ferric oxide	22–55%
Silica	5–18%
Alumina	15–25%
Titania	12–20%
Alkali	3–9%
Loss on ignition	5–9%
Particle size	2–3%

also has been reported to recover caustic from red mud [27]. XRD, XRF, FTIR and thermal analysis of red mud have also been reported in literature [28, 29].

Ordinary Bayer's process and conventional red-mud precipitation technique gives rise to large alkali content in red-mud. A low temperature digestion process (ILTD) [30] has been proposed whereby red-mud is precipitated with alkali content as low as 1.5–2.5%. Processes has also been reported [31] with very short digestion time and relatively low temperature

(135°C) in a tube digester which results in precipitation of red-mud with low alkali contents. Also plants emanating high alkali content have been neutralized by various materials such as gypsum, carbon dioxide etc. [32, 33]

While iron occurs as hematite (Fe_2O_3) in most cases, some percentage of limonite ($FeO.OH.nH_2O$), goethite (FeOOH), siderite ($FeCO_3$), and nontronite ($Fe_2O_3.3SiO2.5H_2O$) is also found in the ore. Ttitania (titanium dioxide) occurs as rutile (mostly), anatase, brookite-ilmenite, titanomagnetic, spene and lecoxene. Similarly alumina occurs in red-mud as undigested diaspore, bohemite (AlOOH), gibbsite [Al $(OH)_3$], kaolinite, and sodium-aluminosilicate (mostly). Undigested Boehmite gives rise to more of hydrargillite phase, while natrolite phase allows better digestibility of the ore. Phase changes in bauxite may occur on heating above 700°C. CaO content of red-mud is generally low (of the order of 2–3%) and generally comes from the deliberate external addition of slaked-lime for conversion of caustic soda to inactive sodium carbonate. Lowering calcium and iron content in the raw ore itself (bauxite) by HCl leaching brings down iron content in red-mud to a large extent [34, 35].

Red-mud is a highly environmental polluting agent and its disposal is still difficult proposition especially considering its voluminous production. While its disposal in open field contaminates potable ground water, its disposal to sea adversely affects marine life. Besides heavy metal contents (which are toxic) and very fine iron content in red-mud which is easily carried through along with alkali, free disposal of red mud contributes heavily towards detrimental effect to land and sea lives. Although many alternatives routes like cleaning up red mud mess through microbes, vegetation etc have been studied with aim to neutralize these toxic effects, its enormous volume of generation still does not offer any easy solution to this problem of its disposal. Moreover, with ever increasing transportation charge, moving such voluminous product to a faraway site is also becoming prohibitive. In recent times various smelter plants tested other alternative routes to convert red-mud into finished products which include building products (additive for cement plant, colorings agents for paint works) , producing toner for papers in wood pulp and paper industries, producing iron ore sinters, improving soil structures, extracting rare earth metals, absorbents, coagulants, landfill etc [36–43].

However above industrial use of red-mud suggested in the past, guarantees consumption to only a fraction of its production even if it is economically viable and thus still dose not address its long term solution. A short description of these processes of making industrial products from red mud is included in following few pages.

(a) Building material
The largest application of red-mud explored so far is for preparation of building materials like ferro cement, bricks, aggregates, concrete [44, 45], glass and tiles, as well as some composite materials. Composite materials include mixing red-mud with fly-ash, clay and slags as well as gypsum to increase weather durability. Fundamental studies like sintering behaviour [46, 47], electrochemical impedance study against steel [48, 49] have thrown light on corrosion inhibition by use of red mud in cement and thereby producing structural materials from red mud. Development of slag-based inorganic polymers using red mud (geopolymers), which because of its superior mechanical properties and weather durability, finds number of applications in construction and building material industries [50–52]. Hardening of gypsum with other inorganic materials like red mud has also resulted in production of cheap reasonably strong cementitious material for application in construction of river embankments [53–56]. Such applications in saline water like sea and ocean fronts creates destabilization problem which has been solved by using organic polymers like epoxy polymers and particulate filled polyester composites with red mud [57–60]. Studies have also been carried out in same line of thinking for use of red mud as clay liners to create hydraulic barrier [61–63].

Frames for door and windows made from red mud although meets the strength, are very heavy to transport. Bricks have been prepared by mixing fillers like fly-ash with red-mud. Due to heavy weight of the bricks and consequent large treatment cost, use of such bricks has been limited to about only 20 km radius of its production site. Similarly concrete for load bearing applications (e.g. for roads and pavements) were also prepared with red-mud. Red-mud itself has been converted to aggregates by polymerization followed by sintering. Artificial marbles were also prepared by mixing a portion of the red-mud with fluid bed boiler coal ash, sintering, glazing and final firing at high temperature. So far one of the successful utilization of red mud in this kind of industry has been in the production of cement or cement clinkers from red mud [64–72].

Coatings for concrete structure were also made by mixing red-mud with bituminous tar. In cement making, addition of red-mud is restricted to 1–5% as further addition found to deteriorate quality of the cement. Higher fraction of red-mud in cement also found to have adverse effect on its durability. Red mud having larger size particles (especially with bauxites having higher silica content) have been used as constructional material for road making [73].

Table 2.11 Typical composition of Portland cement

Constituent	Range
Ferric oxide	2.5–3.5%
Silica	21–23%
Alumina	4.5–6%
Lime	64–66%
Magnesium oxide	1–3%

   Commercial trial related use of red mud in these areas so far includes use of red mud in cement industry in India to the extent of 2.5 million tones in 2006, and as residual bauxite raw material in Japan and Greece. In 2002 bauxite residue (dry red mud) from Alcoa's Kwinana (Western Australia) plant was used to build homes in south-west western Australia. However health department put objection to such constructions as these houses found to emanate radioactivity to an unacceptable limit. Nevertheless, all the constructional materials mentioned above are capable of consuming only a small fraction of red-mud produced. Further, substitution of conventional raw materials with red-mud in industrial products sometime tells upon its quality. Simultaneously long-term durability of red-mud-based constructional materials is still not known. Their extensive use in a localized area may possess health hazard with deterioration and rejection of these product in large scale in near future.

(b)   Extraction of valuable chemicals and metals
   General composition of red- mud indicated in Table 2.8 indicates that products that can be extracted from red-mud are iron, aluminum, titanium and some rare-earths and alkali. Technically although it is possible to extract iron from red-mud, still it can not compete economically to cheap iron-ore / blast-furnace route of conventional iron making. Alumina can be solubilized as aluminum-sulfate by sulfur-dioxide and sulfuric-acid. Titania can be separated by fusing with sodium-bisulfate at 300°C (1:4 ratio) followed by dissolving in water. Ferric-oxide can be separated by Zimmermann-Reinhardt reagent where the ferric iron is first converted to ferrous iron by stannous-chloride, thus making it water soluble. Excess stannous-chloride is later removed by mercuric-chloride. Also acid solution (in sulfuric-acid) when treated with iron filling, converts ferric iron to ferrous iron. Accordingly, Desai & Mohammed [74] treated the red-mud with 10 N NaOH for 8 hours, filtered and the partly alumina free residue after drying, treated with concentrated sulfuric-acid at 130°C for 8 hours, and next ferric ion is reduced by addition of iron fillings (gives

purple colour to the solution). The resultant solution then filtered and heated at 100°C for 1 hour to precipitate titanium as H_2TiO_3. Titania has also been extracted in acid medium [75], for example Swarup & Sharma [76] treated red-mud with sulfuric acid (sp.gr 1.6) and steam was passed through the solution till its sp.gr reached 1.1. The product was filtered and residue was washed and further treated with concentrated sulfuric acid (sp.gr 1.8). This treatment results in the formation of $Ti_2(SO_4)_3$. The product appears like soft cake and was further treated with sulfuric acid (sp.gr 1.6) and again steam passed to attain a temperature of 60°C and specific gravity brought to 1.2. Next ferric iron was reduced by adding scarp iron to the clear solution and $Ti_2(SO_4)_3$ was hydrolyzed by seeding with freshly prepared $Ti(OH)_4$. The residue was treated with coke, in presence of alkali, at 1000°C. The product then extracted with 5% solution of NaOH at 98°C, resulting in recovery of 94.5% alumina and 93.6% titania. However acid, alkali and carbon reduction methods outlined above have not proved economically attractive for extraction of metal values and chemicals although extraction of titania by HCl-$HClO_4$ (which generates 80% pure titania) has been claimed economical with sludges containing higher percentage of titania. Recovery of metal values from red-mud by way of high temperature (1100–1300°C) reduction with coke, sodium-carbonate, and calcium-hydroxide has been attempted at pilot plant level, but economics of these processes are still not competitive with conventional method of these metal production. Iron has been recovered from red-mud after its extraction in metallic form [77–81]. Alumina also has been extracted from red-mud and return to the Bayer plant by roasting red-mud with alkaline oxides and an alkali carbonate at about 800°C [82]. Titanium dioxide has also been extracted from red-mud by acid leaching process [83, 84]. Rare earth has been extracted from red-mud using strong acids like hydrochloric acid, nitric acid and sulfuric acid [85]. Red mud has been used to contain heavy metal contamination from copper producing plants emanating slags and wastes [86]. While silicon containing polymers have been used to improve flocculation of red mud in Bayer's process [87], red mud itself has been used to prepare coagulant for removal of heavy metals from liquid wastes [88]. Selective fungi have also been tried for removal of aluminum from wastes of Bayer's process [89, 90].

(c) Ingredients for paints and pigments
Ferric-oxide and titania occurring in substantial quantity in some red-mud can be extracted to make paint and pigments. For example [91], annealing with 4% sodium chloride at 700–800°C followed by extraction with 1% HCl gives a red-brown pigment with high colouring and covering power, suitable as anticorrosive paint. Sometime barium and zinc chrome in

red-mud gives the paint better covering power than conventional red-oxide or red-lead primer, while barium-potassium chromate enhances its anticorrosive property to a large extent. Similarly addition of vermiculite powder to the above composition enhances its weather resistant property, flexibility and fire-resistant property of the paint. Weather durability of the paint is further enhanced by adding dehydrated castor-oil and cashew nut shell liquid as fluid vehicle. But if a rapid drying formulation is required, above fluid vehicle is replaced by alkyl resins. Yet in some formulations alkaline solution of C_{8-29} carboxylic acids or di- and poly-carboxylic acid half ester with C_{8-23} alcohols are added to produce organophilic layer on the surface of ferric-oxide which work well with conventional recipe involving pure ingredients only.

(d) Other usage:
Among other attempts made to find further usage of red-mud, following processes worth mentioning:

(i) *Production of absorbents for a number of pollutants both in flue gases and waste effluents.* Under this category, particularly effective is the removal of H_2S and SO_2 gases from flue gases. In these formulations dolomite, hydrated lime, sodium hydroxide, sodium carbonate etc are added to red mud in order to increase its absorption efficiency. Similarly activating red-mud by heating at 400–800°C for 30 minutes whereby its specific surface area enlarges by a factor of 3–4, has been used [92] to remove phosphate and heavy metal ions from industrial waste water treatment. Red-mud has also been activated by acid digestion, followed by gelling with alkali for waste water treatment. A composition involving mixture of sand and red-mud has also been developed to remove bacteria from effluent water [93]. Similarly dairy wastes have been treated with a mixture of red-mud and alum to purify the effluent water [94].

(ii) *Catalysts for chemical synthesis.* One such attempt made is to use red-mud to transfer hydrogen from gas to liquid phase in coal liquefaction process. Catalysis effect of red-mud in this chemical conversion reported to be successful to some extent. In such case, ferric-oxide in red-mud appears to be the main ingredient for coal liquefaction process. However thermal stability of red-mud being much higher than corresponding pure compounds, its use as high temperature catalyst has caught attention of scientists for a long time [95–99]. Red-mud has also been tried for demineralization of hydrocarbon oils or heavy waste oils. In these applications, red-mud aided in the removal of metals like nickel, vanadium and non-metals such as sulphur from the hydrocarbon oils. Red-mud has also been tested in gas catalyzed reactions such as, reduction of oxides of nitrogen in flue gases, reduction of sulphur-dioxide gas with carbon monoxide etc. Besides these, red-mud mixed with sulphur has been used towards catalytic hydro liquefaction of biomass at 400°C, and along with sodium-sulphide (neutralizing agent) for hydrogenation/ regeneration of waste oils/used lubricant oils under 150 bar pressure [100]. Red-mud has also been tested as a substitute for conventional drilling mud in drilling fluids [101]. It was found that red-mud was superior as it is more resistant to contaminant ore ingredients (even at 121°C) before precipitation or formation of calcium-silicate. The other advantage of red-mud in this respect is that it has higher ohmic resistance (of the order of 2 ohm meter) as compared to 0.2 ohm meter for conventional lime-based muds. It is also less hazardous from handing point of view and also involves lesser processing cost. Acidification of red-mud with sulfuric acid, followed by neutralizing with ammonia gas till it reaches pH 8, and the resultant residue after adjusting its alumina and ferric oxide content, was made into a flux for refining steel and iron [102]. Red-mud has also been tested for desulphurizing and dephosphorizing molten steel. Red-mud when used as binder in iron as agglomerating charge resulted in increase in strength of green briquettes and increased gas permeability of the charge. While it can be used at a rate 1% in the recipe for binding ores, up to 5% red-mud has been tried in blast furnace sinter mix. Rare earths, like Cb, Sc, Zr, Nb, etc, have been extracted from red-mud using electro-thermal reduction or by leaching with H_2SO_3 followed by solvent extraction with 2-ethyl-hexyl phosphoric acid mono-2-ethylhexyl ester [103]. Ferro cement generated from red-mud has been used for bottom sealing and dam construction. Besides all these red-mud has also been tried as a fertilizer (after neutralizing with gypsum), filler material for plastics (PVC mixed with 25% red-mud and extruded into pipes) which has been found to be resistant to ageing. Red mud in combination with fly-ash has been used to prepare ceramic tiles and dinnerware. Another use of red mud is in producing catalytic effect in methane formation [105–108]. This observation has been used to generate hydrogen and other valuable chemicals from methane using red mud. However besides all these trials, Parekh & Goldberg [104] in a comprehensive report on red-mud utilization concluded that presently there is no hope of finding an economically viable route for red-mud utilization.

(iii) *Red-mud has been used to remove arsenic from water* [109, 110]. Red-mud was chemically modified to absorbed arsenate from water

[111–113]. Red mud has also been used to absorb sulfurous gases from various wastes of industries [114, 115] and neutralizing acid sulfate water, sea water, acid mine water, and acidic soil [116–122]. Neutralization of acidic soil with red mud not only restores pH of the soil but also adds valuable nutrients like phosphorus [123–128]. Red mud has been used to absorb dies like methylene blue, malachite green, rhodamine B, etc, from waste water [129, 130].

(iv) *Refractory materials suitable for high-temperature applications have been produced by calcining red mud at about 1000°C and binding it with colloidal silica, colloidal alumina, sodium silicate and sodium aluminate.* These refractory materials are resistant to corrosion by alkaline materials and fluorides at high temperatures. These refractories are suitable for lining electrolytic cell for production of aluminum while cryolites are involved [131]. Ceramic glazes [132] and glass ceramic as decorative tiles in building industries have also been reported [133–135] to be produced from red mud.

(v) *Attempts have been made to produce ceramic products from red-mud.* For example, glass ceramic made from red-mud without using any catalysts [137]. This was possible due to presence of iron oxide in red-mud [137]. Raw materials in the form of free-flowing granules were produced from red-mud for converting later material to ceramic product [138]. Aggarwal et al. [139] published a review for making white ware ceramics from red-mud. Shaped ceramic products also have been produced by mixing red-mud with silica, silicates or similar materials, or with dolomite to the extent of 10–40% by weight (shaping the resultant mixture thus produced was first given shape in green stage and then fired at about 1000°C [140]). Lesser quantities of red-mud was added to various other materials such as fly ash rock quartz, grogs and stones, glass powder etc. to produce various ceramic forms [141–148]. Ceramic coatings on metal surfaces (like copper) have been reported [149] using red mud and plasma spray process. Heavy clay ceramics has also been reported from red-mud [150]. It may be mentioned here that level of Cr, Ni and V in red mud is high (around 1800 ppm, 1000 ppm and 1500 ppm, respectively) compared to 100–200 ppm in clay. These heavy metals which can cause toxicity in humans, can only be immobilized/stabilized in ceramic matrix. However use of red mud in traditional ceramics has been found to increase natural radioactivity. Radiation dose criteria established with various products restricts use of red mud as follows: in bricks – not more than 14 wt%, in floor and roofing tiles – no restriction (as is the case for use in road pavements), in ceramics for exterior application – up to 61 wt% [151].

Red mud is generated by the processing plant as slurry have pH 12–13. After sun drying it becomes semi-dry and in India this semi-dry material is disposed in barren waste lands. In Greece they are dried and made in to form a cake by high pressure filter press for either dry disposal or utilization by down stream industries. Sintering behavior of red mud in various atmosphere has been studied in detail by Pontikes et al. [152] at University of Patras, Greece in order to exploit the fluxing action of iron in Fe^{+2} state. These studies bear through in sight into the possibilities of forming high quality traditional ceramics from red mud. It may be noted here that morphologically red mud contains alumina in the form of diaspore (Al_2O_3, H_2O), gibbsite(Al_2O_3, $3H_2O$) and other phases like perovskite $CaTiO_3$, calcium alumino iron silicate in the form $Ca_3AlFe(SiO_4)(OH)_8$, cancrinite $Na_3Ca_2Al_6Si_6O_{24}(CO_3).2H_2O$ and possibly geolite $FeO(OH)$ and sodium alumino silicate hydrate [$Na2O.Al_2O_3.1.68$ $SiO_2.1.73$ H_2O], $CaCO_3$ and quartz (SiO_2). Nevertheless, if the major phases are taken as $CaO-SiO_2-FeO$ or $SiO_2-FeO-Al_2O_3$ the triaxial phase diagrams with corresponding eutectics are shown in Figs. 2.1 and 2.2 below. On heating in reducing atmosphere (e.g. H_2), Fe_2O_3 generally first forms FeO. Fe_2O_3 which below 450°C releases metallic Fe. Between 450 to 570°C it further breaks down to $Fe_{(1-x)}O$ which ultimately yields

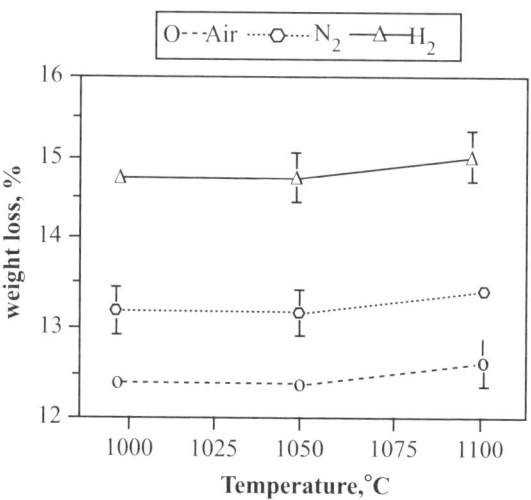

2.1 Dilatometric studies for red mud in air, nitrogen and hydrogen Atmosphere [152].

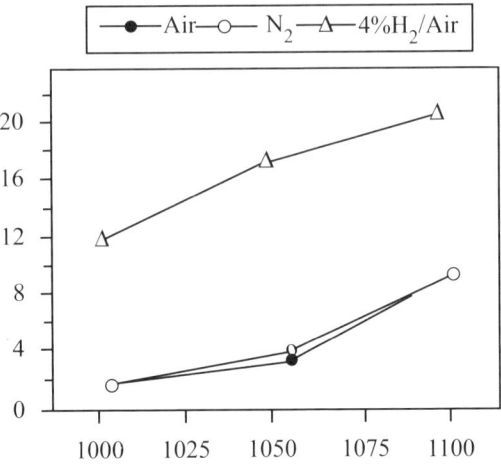

2.2 Shrinkage characteristics of red mud under air, H_2 and N_2 atmosphere [152].

metallic iron. Above 570°C it directly convert to $Fe_{(1-x)}O$ and ultimately to metallic iron. Dilatometric studies done by above authors upto 1000°C and between 1000 and 1100°C in presence of air, nitrogen and hydrogen (4%) argon mixtures are shown in Figs. 2.2, 2.3 and 2.4. All the figures indicate shrinkage of red-mud body (pressed at 39MPa and processed with 50% clay) starts at 250°C in air and N_2 and continuous up to 1000°C. Between 710°C and 1000°C in presence of N_2 there is a sharp decrease in volume due to formation of nepheline [$Na_6(AlSiO_4)_6$], $Ca_3Al_2O_6$ and $FeTiO_2$ from original ingredients (CaO, SiO_2, TiO_2 and Fe_2O_3). Between 1000 and 1100°C there is the formation of spinel [Fe_2TiO_4] from haematite and ilmenite. If the atmosphere is changed to N_2, at high temperature (1000°C) magnetite Fe_2O_3 is formed. Whereas in presence of H_2, the magnetite forms wustite at 1050°C. The red mud sample also becomes more porous (bigger pore size) in presence of H_2 than in air or nitrogen. High temperature firing also develops glassy phase (above 1000°C) and in developing glassy phase iron oxide (hematite and magnetite) works as a filler and "wustite" as flux. Aluminina also works as a filler while CaO works as mineraliser.

Firing in reducing atmosphere (e.g. propane furnace) increases the mechanical strength of the fired red mud body and consequently decreases water absorption capacity in the resultant ceramic.

(vi) Ceramic tiles have been produced both from red-mud [133, 134] and spent pot liner, both are wastes of aluminum industries. In manufacturing ceramic tiles from red-mud water absorption is an important property and due consideration given to this factor during its manufacturer [153]. Tiles having water absorption value to above 4% was produced by firing a mixture of 30%

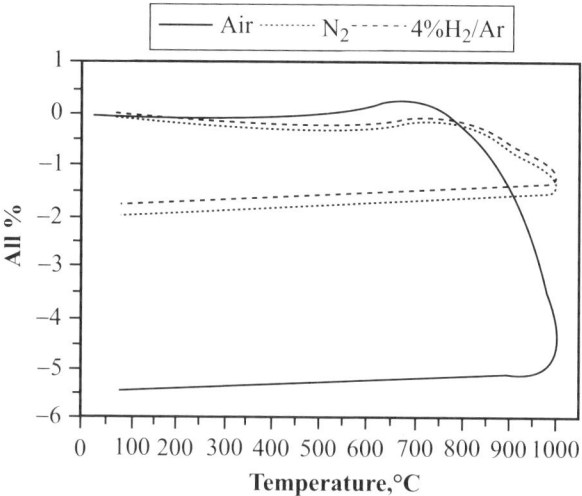

2.3 Linear expansion of red mud under air, nitrogen & hydrogen atmosphere [152].

red-mud, 50% clay and 20% nepheline at 1050°C. Firing at higher temperature brings down water absorption property of the ceramics. Best glazed tiles were produced from a mixture of 20% red-mud, 60% clay and 20% nepheline or perlite. Other additives also have been mixed with red-mud to produce ceramic tiles [154–156, 141]. Ceramic tiles have been produced from spent pot liner after oxidizing the same at high temperature to remove fluoride and cyanide content and burning out carbon value to obtain a vitrified mass which was later mixed with glassformer like soda-lime-silica glass collect, fly ash from utility boilers, incinerated ash lime stone, gypsum, silica sand and nucleating agents like titanium dioxide, zirconium oxide, phosphates, fluorides, etc.

(vii) Other novel uses reported for red mud includes producing radiopaque material [157], production of castings [158], capture of carbon dioxide (greenhouse gases) from atmosphere [159, 160] etc.

2.5   Coal rejects/carbon dusts and coal ash

These wastes are generally produced by captive power plants which are invariably installed within the premises of aluminum smelter plants. Carbon dusts are also collected in anode making plant in the form of dedusting dust and FTA dust. Both sodium and fluoride are the common impurities in these dusts, besides some refractory material may also be associated in FTA dust. Table 2.12 below shows typical sodium and fluoride contamination in these samples.

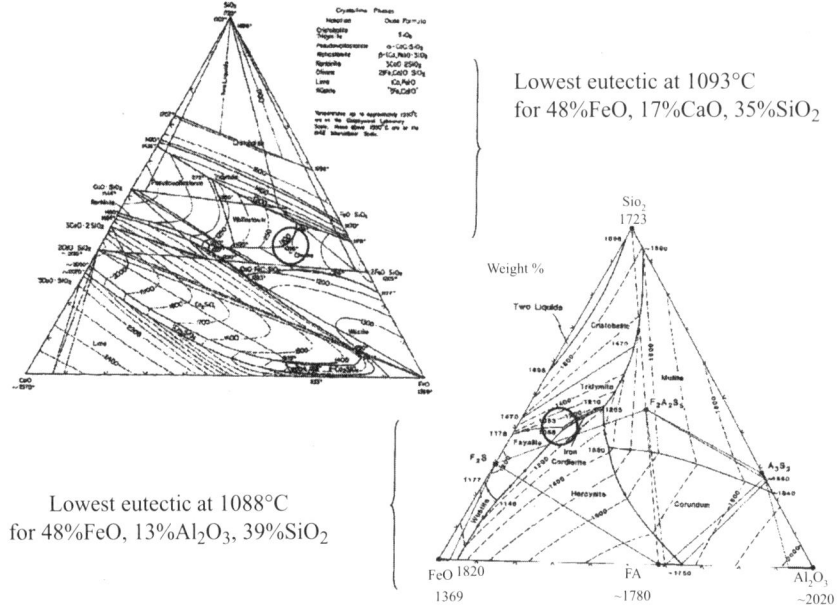

Lowest eutectic at 1093°C
for 48%FeO, 17%CaO, 35%SiO$_2$

Lowest eutectic at 1088°C
for 48%FeO, 13%Al$_2$O$_3$, 39%SiO$_2$

2.4 and *2.5* Eutectic formation in the FeO, Al$_2$O$_3$, SiO$_2$ system[152].

Table 2.12 Typical sodium and fluoride value in anode NALCO, India smelter plant dusts

Contaminants	FTA dust	Dedusting dust	Anode butts
Sodium	1.2	3.0	0.60
Fluoride (%)	0.01%	0.05 %	2–3 %

Coals received at the captive power plants of aluminum smelter unit are first sorted out then sent to crusher plant and finally to milling plant to generate proper size of the coals suitable for boiler use. The sized coals are then fed to the boiler plant for generating necessary steam and finally electrical power for the smelter plant. Coal brunt in the boiler produces ash at the bottom of the boiler known as bottom ash and fly ashes carried over by flue are arrested either by cyclone or electrostatic precipitator and finally the flue is scrubbed with water. All these ashes mixed together are hydraulically transported to the ash pond and settled there. These ashes create environmental problem and its use as well as transportation and disposal is a long standing problem for all captive power plant operators. Physical and chemical properties of the coal used depend on the mine from where they are procured and consequently composition of ashes varies from plant to plant. In India majority of these coals have low volatile, high ash content. Table 2.13 below shows typical composition of such coals, along with other coal rejects such as crusher plant dust, mill rejects, and bowl mill outputs.

Table 2.13 Typical composition of carbon rejects from power plant

Material	Moisture	Volatile matter	Ash	Fixed-carbon	Calorific value
Raw Coal	4%	23%	42%	31%	3973 kcal/kg
Crusher plant dust	3.66%	22%	50%	24%	3416 kcal/kg
Mill rejects	3.82%	27%	29%	40%	5251 kcal/kg
Bowl mill output	2.18%	27%	37%	34%	4783 kcal/kg

Table 2.14 shows typical composition range of constituents in fly ash/ bottom ash as well as accumulated pond ash in the captive power plants of aluminum smelter plants.

Table 2.14 Typical composition range of fly ash, bottom ash and pond ash

Components	Fly ash / Bottom ash	Pond ash
Alumina	28–33%	22%
Silica	48–60%	63%
Titania	2–3%	1.5%
Ferric-oxide	3–4%	4%
Loss on ignition	0.5–1.0%	7%

All the carbon rejects/dusts mentioned above are high in ash content and can be burnt by conventional chain-grate burner. Use of these ash coals and rejects will be discussed in detail in next chapter..

3

Treatment of the wastes and quality of byproducts

3.1   Treatment of spent pot liner

Usual contaminants in spent pot liners have been mentioned in previous chapter. Attempts have been made to decontaminate these spent pot liners by hydrothermal and alkali leaching techniques. While hydrothermal route is not effective in removing all contaminants to desired level, alkali leaching removes most of the contaminants but requires too long time and demands fine powdering of the spent pot liner. Accordingly, the general shop floor practice has been to powder these carbon blocks and burn them in a pulverized fuel burner [161–166]. Free burning of SPL by this means requires gas scrubbing to avoid air pollution, and valuable carbon is lost by this process and helps recovering calorific value only.

As mentioned earlier, pot liner are laid on insulating and fire bricks and electrically connected through a bus-bar made of iron. The waste spent pot liner is taken out of the cells, in three batches – the first batch, the top layer of the spent pot liner having no refractory contamination, the second batch contaminated to some extent by refractory bricks, and the third batch contains mostly brick chunks and iron contaminants.

Attempts have been made to separate carbon from mineral matters either by physical method or by chemical means. Chemical method of treating SPL has been described in detail in later section of this chapter, while physical method of liberating mineral matter from SPL will be discussed in following pages. Fernandez et al. [167] worked on recovery of carbon powder from SPL for anode production. Figure 3.1 below shows the distribution of inorganic fraction and carbon achieved by crushing first cut batch of SPL and subsequent screening by above authors.

The inorganic fraction of SPL was concentrated in the smallest size reflecting the fact that inorganics are present in SPL as smaller chunks (as can be seen in Figs. 3.1 and 3.2 below).

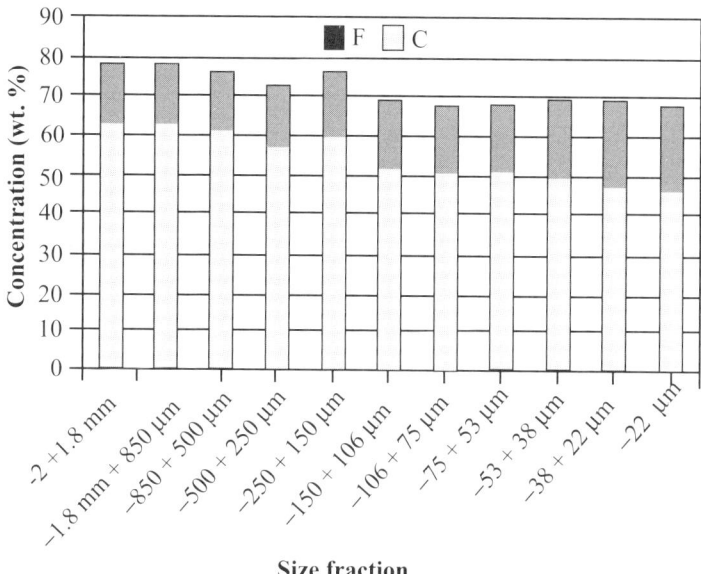

Size fraction

3.1 Fluoride and carbon distributions in SPL size fractions.

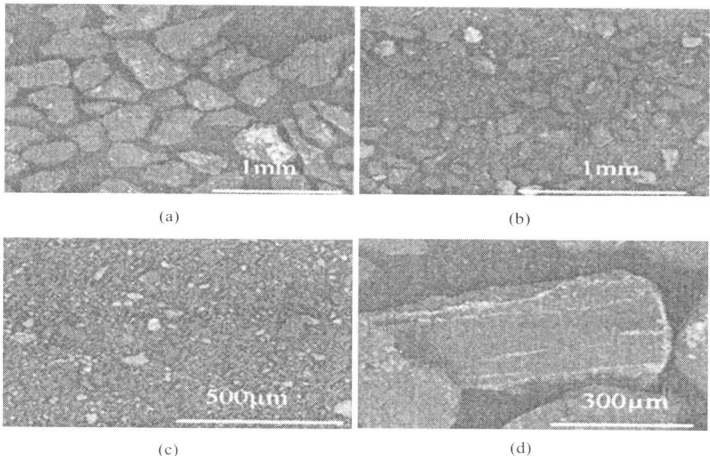

3.2 Comparison of mineral liberation within different SPL size fraction (a) −500+250 micrometer, (b) −150+106 micrometer, (c) −53 micrometer, (d) section of SPL particle.

Grindability therefore plays an important role in separating mineral matter carbon in SPL although difference between 62.75% and 45.8% is not great. XRD diffraction pattern of inorganic matter separated from SPL by ashing at 800°C for hours by above authors are shown in Fig. 3.3 below.

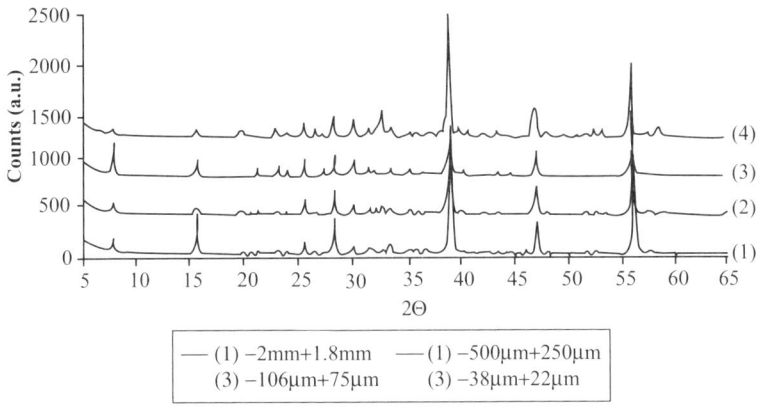

3.3 X-ray diffraction patterns inorganic fraction from selected SPL size fractions.

Different size fractions are compared in above figures, major peaks correspond to NaF, Na_3AlF_6, CaF_2 and $NaAl_{11}O_{17}$. For solids recovered from the −2 mm to +1.8 mm fraction, peaks at 38.96° and 56.15° 2 corresponds to NaF whereas peaks at 28.30° and 47.105° 2 are essentially CaF_2. $NaAl_{11}O_{17}$ peaks at 7.90°, 15.80° and 33.50° 2. Na_3AlF_6 can be identified for signals at 22.85, 32.55 and 46.70 2 in patterns 4 in above figure, (−38 + 22 µm fraction). As we all know fine grinding of hard material especially material like SPL, consumes lots of power in mills, added to this fact as shown above, mineral liberation from SPL by grinding is not complete. These facts have given rise to attempts by research workers to treat SPL by chemical means in order to completely separate carbon from the mineral matters.

Management of highly toxic SPL (categorized as K088 under hazardous wastes) is reported in literature and can be categorized into four groups:

(a)   Producing industrial materials directly from SPL while decontaminating toxic components.
(b)   Separation of valuable chemicals (like fluorides) and recovery of carbon value.
(c)   Direct use of SPL after diluting with inert materials (like sand and lime) and use as landfill.
(d)   Utilization of its calorific value by direct burning and scrubbing the flue gas to avoid air pollution.

We will discuss these methods in nutshell here citing related references. Direct treatment of SPL at high temperature generally intended for destroying the fluorides and cyanides and sometime the molten mass is drawn to generate inorganic refractory fiber, an useful industrial product. Inerting the SPL by dilution with silica, lime etc is also carried out at high temperature resulting in production of a less harmful product. In one such attempt by Sims et al.

[168] SPL mixed SPL with other inert materials such as silica, lime etc. and subjected to a temperature of 1600–3100°F. While the fluorides and cyanides destroyed by this means the molten material was suitable for drawing vitreous fibers. These minerals wools find a number of high-temperature applications in industries. In a similar attempt Morgenthaler et al. [169] at Columbia Ventures Corporation, Vancouver treated SPL in a plasma torch furnace at a temperature of about 1000°C. At this temperature carbon is gasified and converted to combustible carbon monoxide, hydrocarbons and carbon dioxide. Fluoride component is reduced to gaseous HF and cyanide components are destroyed. The treated SPL thus becomes non hazardous and can be disposed in open field. Simultaneously, gaseous components like fluorides were recovered by a condenser. Similarly, Banker at al. [170] detoxified SPL at about 1200–1700°F by mixing it with lime stone. O' Connor at al. [171] mixed iron ore with SPL and the mixture was treated in an electric arc furnace at about 1350–1650°C. The process, while destroying fluorides and cyanides, liberates molten iron metal from its ore. Further the off gas was treated in a scrubber using sodium hydroxide to capture hydrofluoric acid gas as sodium fluoride. In a similar attempt Davis et al. [172] treated SPL at about 1000–1700°C. The liquid slag was held at elevated temperature until at least a portion of the contaminating components have either decomposed or evolved from the melt as a gas. The slag is subjected to cooling and the contaminating compounds are bound or encapsulated in to a solid glassy slag. It is advantageous to add silica to such waste as it forms a glassy solid sodium metal silicate matrix encapsulating fluoride residue in the solid matrix. Most of the sodium simultaneously get fixed in the silicate matrix.

Output of above high temperature furnaces are reported to convert SPL Aluminum Company and Alcoa in USA use these inert materials for land filling. However other aluminum smelters are critical of the processes because it results in an increase in volume of waste for land fill, the transportation cost for shipping the SPL to the treatment plant which exceeds the treatment cost themselves and not meeting the treatment standards adopted by EPA (TCLP). Moreover, this residue generally has a final value which is comparable to the volume of the input. Pyrohydrolysis and fluidized bed reactors [173–175] on the other hand produce a mass which allegedly contain non-leachable material as it is embedded in a glassy phase. The Pyrometallurgical process adopted by Portland Aluminum (Victoria, Australia) utilized carbon content of SPL and supplements natural gas for the caloric value in SPL [176]. Pyrohydrolysis involves contacting SPL with water or steam at high temperatures, whereby the water introduced reacts with the fluoride compounds to form HF. However, it has been found that while the pyrohydrolysis of aluminum fluoride is relatively easy, calcium fluoride and particularly sodium fluoride are difficult to react [177, 178]. However, the process described in the aforementioned patents requires exceptionally high temperature and excessive quantities of steam. In a more recent patent to Martine Marietta Corporation [179] there is described a pyrosulpholysis procedure for the treatment of SPL which involves high temperature treatment of SPL with air, steam and sulfur dioxide in a single reactor. The reactor may be a fluidized bed, packed bed or close furnace. The recovery process involves the following reactions in the waste material (a) Decomposition of cyanide, (b) combustion of carbonaceous and hydrocarbon material, (c) oxidation of sulfides and nitrides, (d) sulpholysis of the fluoride salts there by forming HF gas and sulphate salts. A major disadvantage of using a single reactor is that all reactions take place in one reactor with reactants necessarily present in low concentration which is not conducive to high reaction efficiency. The conversion of sulfur containing reactants particularly tends to create emissions which constitute environmental hazards in addition to those costs by fluorides and cyanides. The variable chemical composition of the feed stocks makes control even more difficult in a single reactor. These short comings were overcome by conducting sulpholysis in a separate stage [180]. In another patent granted to Aluminum Pechney [181] SPL of size <5 mm was mixed with calcium sulphate and injected in to the center of a vortex formed by a flow of circulating hot gas arriving tangentially at the top part of a reactor having temperature between 700 and 1100°C and having a whirling movement.

Attempts have also been made to stabilize toxic elements in SPL through cement formation. Here SPL has been used either as an alternate fuel or as a direct substitute of coke in cement manufacturer. SPL's average fuel value is approximately 2/3 that of coal, half the equivalent of coke and about twice that of refuse derives fuel. Use rates of up to 10% total fuel requirement were reported by Giant Cement Company. Here use limitations were reported as being the sodium concentration. Introduction of processed SPL into the preheater portion of the cement kiln system, at specific temperatures are the basis for the CIEC's patent claim for NOx reduction. A temperature window of 1400–1600°F appears to be the optimum temperature for NOx reduction from the ammonia/cyanide components. These temperatures are found within the preheater tower section of the kiln, with the exact location determiner based on pre heater tower design. Formally introduction would be via gravity or mechanically placement as opposed to use of pneumatics which would introduce tramp air and potentially reduced production rate. Once introduced no further handling or processing is needed. Immediately after to SPL introduction and ignition a small amount of water would be added to accelerate the evolution of ammonia and cyanides.

At the Federal University of Para (Brazil), Silveira at al. [182] stabilized toxic elements in SPL by cement based systems. The studies were carried out with concrete hexagonal blocks manufacture with a constant mass of 10% of the waste, 20% cement, and varied % of water, coarse aggregate, sand and other additives. Porosity and compressive strength of the concrete masses were controlled by using micro silica and super plasticizer. Results showed and average pH value for the SPL inorganic fraction and fragmented blocks as 10.2 and 11.1, respectively. Effectiveness for control of leachable cyanides and fluorides by this process were 59.33% and 57.95%, respectively.

In order to recover fluoride chemicals and carbon value from SPL both acidic and basic treatment of the wastes have been carried out by various researchers. While heating at lower temperature of about 450°C and mixing the heated SPL, with water produces reaction gases and residue for recovery of chemical and carbon from SPL most of the chemical recovery processes attempted are through direct acid or alkali leaching of SPL. In one of the former process, Cooper at al. [183] treated SPL in a rotary kiln in to which air was introduced at about 450°C. Cyanide was destroyed at about 750–800°C and by holding SPL for about 40 minutes at that temperature.

Wet methods of SPL treatment reported in the literature may be categorized under three headings: (a) treatment with neutral solvent like water, (b) treatment with alkali like sodium hydroxide, calcium hydroxide etc, (c) treatment with acids like sulfuric acid, nitric acid, perchloric acid, chromic acid, etc. Water dissolves sodium salts like sodium fluoride from SPL. SPL to water ratio is generally maintained at about 1:3 to 1:8 ratio, at a temperature of 20–70°C for about 10–20 minutes [184]. The process requires fine grinding of SPL to −48 Tyler mesh. During leaching process there is a continuous reprecipitation of sodium aluminum silicate compounds on the surface of the particles. These compounds tend to block the pores of SPL particles and stops solubility of residual fluorides. They also tend to inhibit the reaction of caustic solution with cryolite track inside the pours. Accordingly, to assure affective leaching it is important to remove this coating in a continuous manner. This is achieved by acid activation which removes these obstructions without significant process or economic penalty. Generally water leaching is followed by alkali leaching to enhance solubility of fluoride compounds. Interestingly alkali (sodium hydroxide and potassium hydroxide) reaction followed by heat treatment up to 730°C results in activation of the residual carbon [185, 186]. Alkali reaction generally requires grinding SPL to size <100 microns to ensure complete extraction of soluble fluorides. Alkali leaching is capable of removing approximately 55% of fluoride available in SPL. It may be noted here that SPL when left in the open, turns grey because of hydrolysis of unreacted cryolite with moisture present in air forming sodium hydroxide which brings about above reaction. Alkali reaction precipitates fluoride in SPL as AlF_2OH which subsequently can be converted to AlF_3. In above process SPL can be leached by counter current extraction with a caustic solution having a concentration of about 14 g/l sodium hydroxide [187]. Lime also reacts with pot lining chemicals and precipitates as fine calcium fluoride salts. Calcium fluoride can be used as metallurgical spar or it can be treated to form hydrogen fluoride and aluminum fluoride. The lime leach produces a solution containing dissolved alumina in dilute caustic which can be returned to the Bayer plant. In such process crushed SPL is ground in presence of a quantity of lime stoitiometrically adequate to combine with all fluoride there in and forming an aqueous suspension of said fluoride and lime which is kept in an agitated state for sufficient time to bring about the precipitation of calcium fluoride with the release of soda [188]. Heating the aqueous suspension to a temperature above 140°C in presence of an adequate quantity of clay fixes the free soda through formation of an insoluble synthetic silicate compound. Clays like Illite, kaolinite and smectites are used for above purpose. A pilot plant using above low caustic leach and lime process (LCL & L) is run by ALCAN at RioTinyo, Quebec, Canada. It may be noted here that hygroscopic CaF_2 is difficult to precipitate as "acid grade" and its requirement in aluminum smelter plant is very small. In fact, fluoride is needed in aluminum smelting both as Na_3AlF_6 and AlF_3. While low value Na_3AlF_6 is being produced in aluminum smelter, high value AlF_3 must be added continuously to maintain fluoride balance in the cell. Electrolysis cell design determines AlF_3 product specification as smelter grade (SG). SG AlF_3 production from SPL has been found feasible by Ausmelt-Alcoa (Australia) thermal process. In this later method, incorporating SIROSMELT technology (commercialized as the Ausmelt Technology), the treatment system comprises two major developments. Firstly, after suitable preparation, the spent pot lining is fed into an Ausmelt-designed top submerged lance furnace in which the cyanide-forming materials are destroyed at temperatures up to 1250°C and the contained fluorine is driven off as hydrogen fluoride in the off-gases. In the second major step, a unique gas treatment process developed by CSIRO Minerals and Portland Aluminum converts the hydrogen fluoride in the off-gases to aluminum fluoride in a multi-stage fluidized bed reactor. The furnace also produces a granulated slag referred to as 'synthetic sand'. The aluminum fluoride is recycled into the aluminum smelting process and the synthetic sand can be used in commercial applications such as road making and concrete products.

   AlF_3 obtained by acid leaching of SPL is highly stable in aqueous solution but extensive evaporation is needed to recover the chemical. Sodium and calcium promotes formation of Na_3AlF_6, $NaAlF_4$ and CaF_2. Industries produce AlF_3 by following processes:

(a) $CaF2 + H_2SO_4 \longrightarrow 2HF + CaSO_4$

(b) $Al(OH)_3 + 3\,HF \longrightarrow AlF_3 + 3\,H_2O$

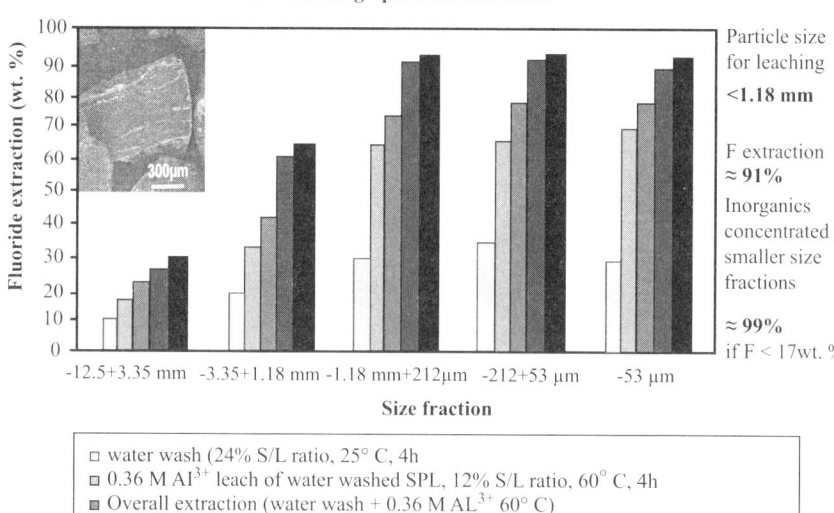

Al³⁺ leaching - particale size effect

3.4 Effect of particle size on leaching from spent pot liner [167].

Production of AlF_3 from AlF_2OH requires milder leaching and reaction conditions (25–70°C) and there is no HF in the gas phase. It also helps in corrosion mitigation:

$$AlF_2OH + HF \rightarrow AlF_3 + H_2O$$
$$AlF_2OH \cdot 1.4\ H_2O + HF \rightarrow AlF_3 + 2.4\ H_2O$$

AlF_2OH on the other hand has been produced by leaching SPL with dilute sulfuric acid in the pH range 0–3 (if needed adding aluminum in an acid soluble form) precipitating AlF_2OH hydrate from the aqueous phase by increasing the temperature to about 90°C and raising the pH 2 to about 4 by continuous and controlled addition of aqueous sodium hydroxide solution [189]. Integration of caustic waste from Bayer aluminum plant for the production of $AlF_2OH \cdot 1.4$ H_2O brings about zero waste approach. Both particle size of SPL, temperature and pH have profound effect on the leaching procedure as can be seen in Figs. 2.4, 2.5 and 2.6 shown below.

SPL has also been treated with calcium chloride/HCl and then with $FeCl_3$ in a single charge to a reactor in order to destroy hazardous cyanide and polynuclear aromatic and to convert it to insoluble floor spar for final disposal. PNAs are destroyed by acidic $FeCl_3$ leach mill solution in an oxygen containing environment at elevated temperature and pressure [190.] SPL is also been directly treated with hydrofluoric acid in order to dissolve all its inorganic matter [191]. The inorganic materials and fluorides in the filtrate are

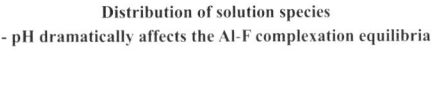

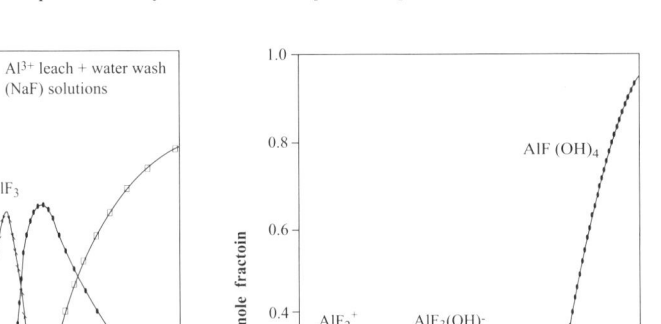

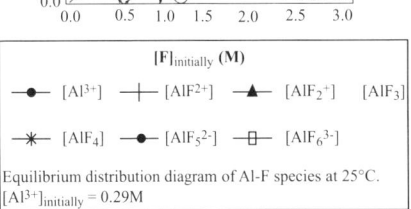

3.5 Effect of pH on extraction of fluoride chemicals from spent pot liner [167].

subsequently separated by precipitance technique. Further treatment of SPL with HF at higher temperature has been claimed to recover fumed silica from spent pot liner [192]. Heating to 400–1000°C forms silicon tetra-fluoride and escapes with HF in gas phase. Silicon tetra-fluoride is subsequently hydrolyzed to fumed silica.

Besides hydrofluoric acid, SPL has been extracted by other oxidizing acids such as nitric acid, sulfuric acid, perchloric acid, chromic acid, etc [193, 198]. These acids have been used either singularly or combination to bring out the desired reaction. Further David Jenkin at Comalco Aluminum Limited (Australia) [193] has first crushed SPL to a size less than 600 micrometer and calcined in a furnace in the temperature range 680–850°C to produce a low cyanide ash which was subsequently extracted by combination of hydrochloric, sulfuric, nitric acid. Average carbon burn out in this experiment was about 58%. In other experiments mentioned above, SPL was directly treated in an acid digester, the off gas contained HF and HCN which can be recovered in appropriate recuperate. Adjusting pH after acid digestion to basic range helps formation of aluminates in the solution while precipitating impurities containing the group calcium, iron and magnesium

Al^{3+} leaching – effect of temperature

$Al(NO_3)_3$ leaching

99.9% fluoride extraction with 0.36M
Al^{3+} in 0.5 HNO_3 at 60° C

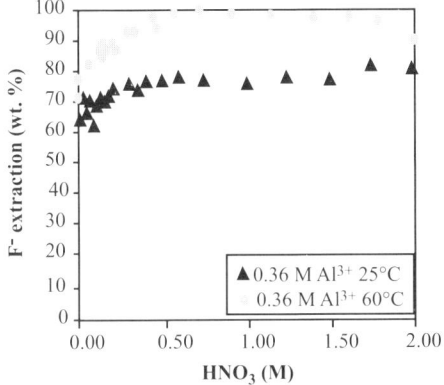

Leaching with Al^{3+} waste

Leaching water washed SPL with Al^{3+} waste:
70% F⁻ extraction (all *leachable* F⁻)
Typical al^{3+}
waste conc.

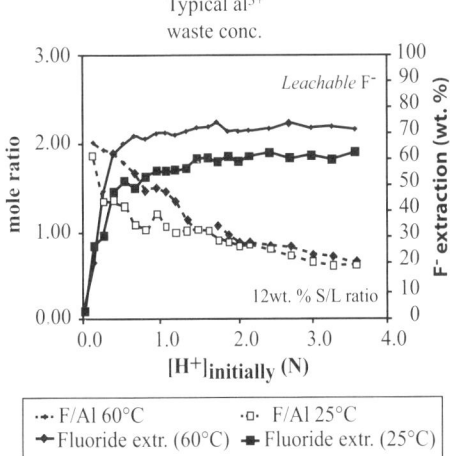

3.6 Effect of temperature on leaching of
salts from water extract of SPL [167].

and then filtering the impurities from liquid. Adjusting the pH in the filtrate helps precipitating alumina trihydrate. The digester temperature is generally held between 135 and 300°C. It may be noted that recovery of fluoride as aluminum fluoride from SPL by alkali leaching leads to precipitation of cryolite until special and rather complicated measures are taken. Similarly, recovery of aluminum fluoride by precipitation method involving acid, presence of foreign ion (i.e. other than aluminum and fluorine) complicates the process and sodium along with aluminum forms double salt (e.g. chiolite and/or cryolite) instead of aluminum-fluoride, the desired product.

The author of this book improvised a chemical technique which while removing its toxic contaminants in shortest possible time, also provides a means for recovering its carbon value for possible use by downstream industries. We will discuss this new process at length in this section. As mentioned earlier the molten salts (alkali and fluoride) penetrate through interstices of carbon blocks forming compound with it and with increase in molten volume generates internal pressure and helps propagation of the crack. Accordingly more and more salt react deep into the carbon block finally results in failure of the pot liner. For this reason, a strong chemical oxidation route was chosen by the present author to remove the contaminants by breaking the already formed carbon-contaminant bond. One such chemical investigation was chromic acid [199]. Reaction of freshly prepared chromic acid with spent pot liner contaminants is instanous and highly exothermic. Accordingly, the reaction rate and consequent exothermic heat can be controlled by choosing appropriate particle size of SPL. For example, while with −60 mesh spent pot liner powder the reaction is instantaneous and gives rise to reaction heat as high as 400°C, the reaction time is around 15–20 minutes only with half inch size particles of the spent pot liner. This in turn provides an inherent advantage to the process, not requiring excessive fine grinding of hard and large carbon blocks of spent pot liner.

The reaction being highly exothermic, with finer particle size it tends to burn out amorphous carbons present in the spent pot liner with high reaction heat; but the graphite particles present in the mixture are spared from such oxidation reaction. As mentioned earlier, since most manufacturers use around 30% of graphite in their recipe for making pot liner, corresponding yield with above drastic oxidation process have been found to be around 30% with same SPL. However if a less severe oxidation step is chosen (for example with perchloric acid) [200] the yield may go up as high as 85% even with finer particles but here the yield will be a combination of amorphous and crystalline (graphite) carbon particles. Accordingly the process provides the flexibility of recovering both crystalline (graphitic) carbon or a mixture of amorphous and crystalline carbon according to the severity of oxidation reaction chosen. In this process chromic acid is being produced by standard method of mixing potassium dichromate with concentrated sulfuric acid. Completion of the reaction is self indicative as bubbling of gases ceases as the reaction goes towards completion. In such reaction all contaminants find their way either in gas phase (which can be absorbed by dilute alkali like sodium-hydroxide) or in the filtrate and washing water. Since the drastic oxidation attacks amorphous carbon, carbon with short range order remains unburnt and thus the product is a micron size powder irrespective of the size of the starting material. Figure 3.7 below shows the particle size distribution of the carbon powder

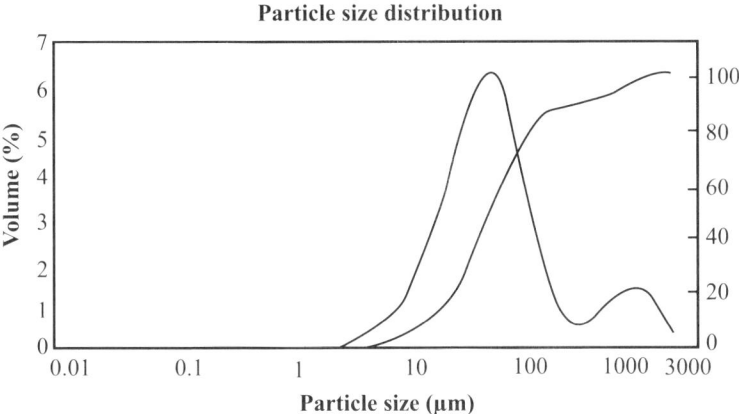

3.7 Malvern Particle size distribution of graphite powder obtained after processing ~ 100BS mesh spent pot li ner powder.

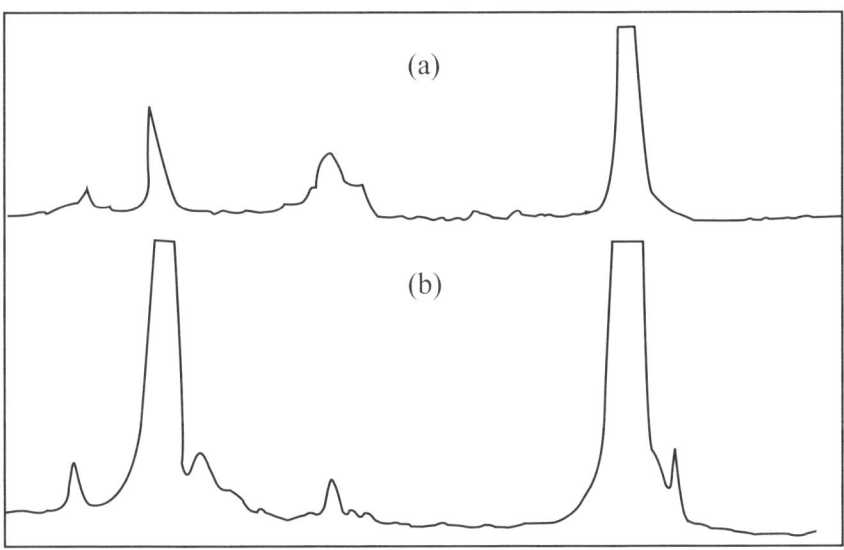

3.8 Comparison of XRD graph of derived carbon powder and natural graphite.

obtained with −100 BS size SPL powder. As the figures show, average particle size of the carbon recovered is around 20 micrometer with specific surface area of about 0.3998 sq m/cc.

While the product qualitatively confirms presence of graphite with shining metallic appearance on rubbing between fingers, instrumentally it is confirmed by XRD as shown in Figure 3.8 below. Table 3.1 shows typical results of oxidation treatment.

Table 3.1 Results of acid treatment of two different grade spent pot liners

Sieve size of the spent pot liner	Fluoride before treatment	Alkali before treatment	Free Al before treatment	Fluoride after treatment	Alkali after treatment	Free Al after treatment
0.25 inch to dust	8%	7	2	0.0132%	0.49%	Nil
− 100 BS mesh	8%	7	2	0.0039%	0.40%	Nil

Table 3.2 Characteristics of the recovered carbon after chromic acid treatment

Sieve size of the spent pot liner	Av. Particle size of recovered carbon	Yield %	Ash content
0.25 inch to dust	19.98 micrometer	30%	11%
− 100 BS	19.98 micrometer	28%	10%

Ash content in the recovered carbon powder found mostly in the form of aluminum and grits which can be easily separated by froth floatation techniques as their density deference against pure carbon and inorganic contents are high. Above results have been obtained by the authors in bench scale experiments. As will be seen in following chapters, when the process was repeated in large scale pilot plant level, average particle size of the recovered carbon was somewhat larger than bench scale results mentioned above. Nevertheless, fluoride and cyanide content in the recovered carbon from pilot plant trials were same as above.

The carbon powder obtained after the acid reaction, because of the severity of oxidation, may form oxygen derivative of graphite, known as mellitic acid. Structure of mellitic acid is shown below. Mellitic acid is also known as hexacarboxilic acid, pyromellitic acid, trimellitic acid, and terephthalic acid. The hexabasic acid derives its name from the occurrence of aluminum salt $(Al_2C_{12}O_{12}, 18 \ H_2O)$, as the mineral mellite (honeystone) found in brown coal. Mellitic acid has a melting point of 288°C (in sealed tube) and the hexamethyl ester has melting point of 188°C.

Molecular weight = 342.12
C = 42.12%, H = 1.77%, O = 56.11%

3.9 Structure of mellitic acid.

Melting point = 286–288°C in sealed tube with decomposition.
Solubility = soluble in water, alcohol, boiling concentrated sulfuric acid.

Sealed tube mellitic acid yields a stable trianhydride that sublimes when heated at 200°C at 3–4 mm pressure. Acid dissociation constant of mellitic acid are pK_1 = 1.40, pK_2 = 2.10. pK_3 = 3.3, pK_4 = 4.8, pK_5 = 5.89 and pK_6 = 6.96. The mellitic acid formed in present oxidation reaction can be undone, back to graphite, by thermal shock at 900–950°C. The washed carbon after drying at 110°C in an oven is inserted to a 900°C preheated furnace for 1–2 minutes and taken out, which gives a free flowing carbon power.

Filtrate and washing liquid from above process contains chromium salt (in +3 states) and some other contaminants. Chromium (III) is much less harmful than the chromium (VI). However chromium in the waste liquid needs to be treated before discharge of the water in open drain. Indian Bureau of standard sets the limit of chromium (VI) to 0.05 mg/l in public water supply and 0.1 mg/l in surface water. The techniques available for removal of chromium (VI) can be broadly classified into two groups – one where chromium is not recovered and another where chromium is recovered for reuse [201]. In the recovery process the metal is recovered by one of the following process – membrane adsorption, ion exchange method, or by biological route. Membrane process includes elecrtodialysis [202], reverse osmosis, ultrafiltration, nanofiltration, and liquid membrane permeation technique. Because a single pass through electrodialysis cell usually removes 30–60% of the metal, recycling of metal rich water is necessary to the cell, until the metal concentration reaches the desired limit. In liquid membrane permeation technique, chromium is recovered as chromate and water can be recirculated in the process for progressive lessening of metal concentration. In this process chromium (VI) is first immobilized by an organic phase (consisting 94% of white spirit as solvent and 3–8 volume% of trialkyl amine) in an acid medium forming its complex and later releasing the metal selectively in an alkali solution as chromate (Na_2CrO_4). In absorption technique, absorbents like activated carbon, fly ash; rich bran etc has been used.

Destructive method on the other hand uses reduction and/or precipitation technique. Chromium (VI) is reduced to chromium (III) at pH 2–3. Commonly used reducing agents are sodium-bisulphite, sodium-metasulphite, sodium hydrosulphite, sulphur dioxide gas and base metals like iron, aluminum, zinc etc. Reduction with sulphur-dioxide can be represented by the equations:

$$2\ H_2CrO_4 + 3\ SO_2 = Cr_2(SO_4)_3 + 2\ H_2O$$
$$2\ H_2CrO_4 + 3\ Na_2S_2O_5 + 6\ H_2SO_4 = Cr_2(SO_4)_3 + 6\ NaHSO_4 + 7\ H_2O$$

Chromium (VI) can also be reduced to chromium (III) by ferrous-sulfate, followed by neutralization with lime to precipitate the heavy metals:

$$K_2Cr_2O_7 + 6\ FeSO_4 + 8\ H_2SO_4 = KHSO_4 + Cr_2(SO_4)_3 + 3\ Fe_2[SO_4)]_3 + 7\ H_2O$$

However, direct precipitation of the metal as hydroxide or sulphite is the commonly preferred method for waste treatment. This is usually achieved by using flocculating or coagulating agents. Precipitation is being carried out in the pH range of 8–9. Neutralization is commonly done with sodium-hydroxide or calcium hydroxide.

$$Cr_2(SO_4)_3 + 6\ NaOH = 2Cr(OH)_3 + 3\ Na_2SO_4$$
$$Cr_2(SO_4)_3 + 3\ Ca(OH)_2 = 2\ Cr(OH)_3 + 3\ CaSO_4$$

The electrochemical precipitation [203] technique uses six steel plates as anode and cathode operated in a bipolar mode with DC power input. Chromium content of the waste water used was 570–2100 mg/l, and after electrolysis it came down to as low as 0.5 mg/l. The precipitated sludge in the cell after electrolysis found to be 68% ferric-oxide and 25% $FeCr_2O_4$. Pure iron powder is also capable of precipitating Cr (VI) as a combined hydroxide [$(Cr_xFe_{1-x})(OH)_3$]. However it requires longer time of contact [204].

Figure 3.10(a), (b) and (c) below shows the pilot plant and Figure 3.11 schematic diagram of the pilot plant tested for above SPL treatment process with oxidizing acid. Figure 3.12(a), (b), (c) and (d) below shows the characteristics of the carbon powder obtained from this plant. Figure 3.13 shows the floatation column used for recovering ash particles from above carbon content achieved by this froth floatation technique. Method applied for froth floatation of coal [205] was also tried by the authors for separation of ash from SPL derived carbon powder. Here the carbon powder was first sieved through BS mess 60. Then 109 liters of water was added to 8.5 kg SPL derived carbon powder in the floatation column. The mixture was

3.10(a)

3.10(b)

3.10(c)

3.10   (a), (b), and (c) 100 kg/batch SPL processing pilot plant
(NALCO, India).

agitatedly a stirrer revolving at a speed of 900 rpm for about 10 minutes; pH
was measured and controlled to about 10. After 10 minutes 21.25 ml diesel
oil was added and conditioned for about a minute. Next the frother and the
stabilizer (i.e. sodium silicate 28.3 ml and MIBC 21.5 ml) was added and air
from the blower was started. The froth emanating from the top of the column
was collected.

3.2   Treatment of red-mud

Composition of red-mud has been shown in Table 3.14 earlier. As can be
seen from composition of red-mud its largest constituent is ferric oxide.
Some red-mud also contains larger proportion of titanium-dioxide. Such

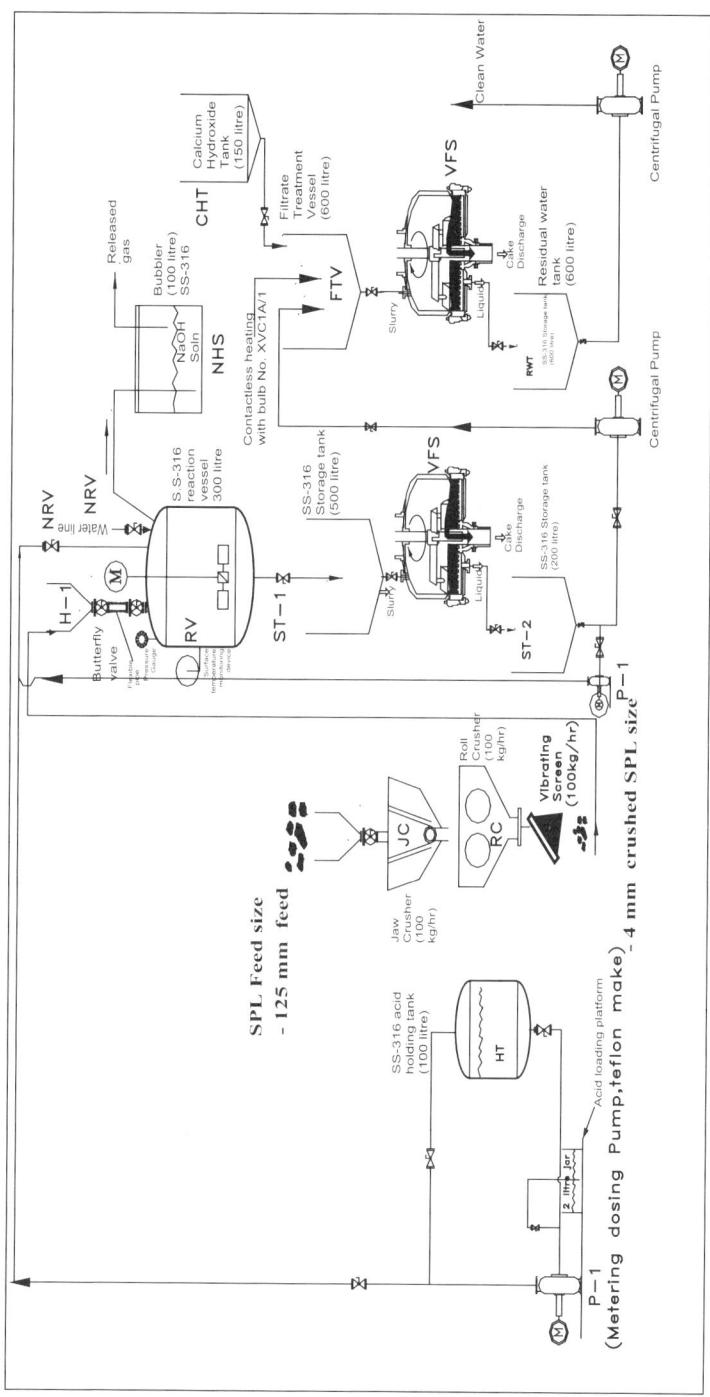

3.11 Schematic diagram of the 100 kg/batch SPL processing plant (NALCO, India).

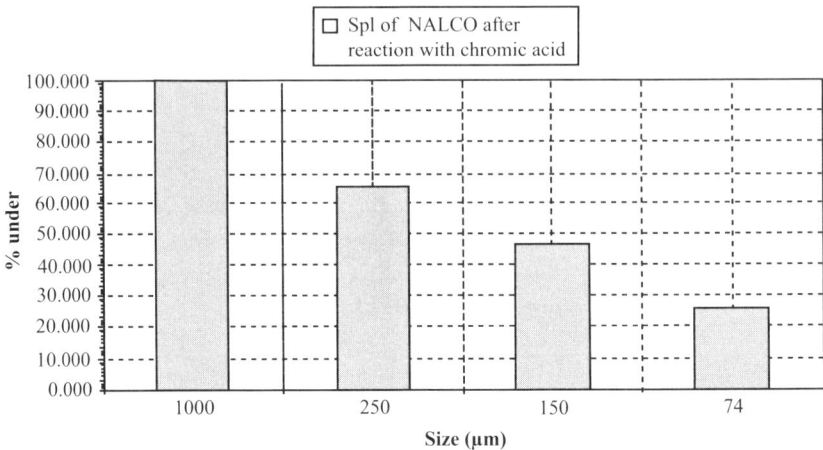

3.12(a) Graph showing d$_{50}$ of SPL after reaction with chromic acid.

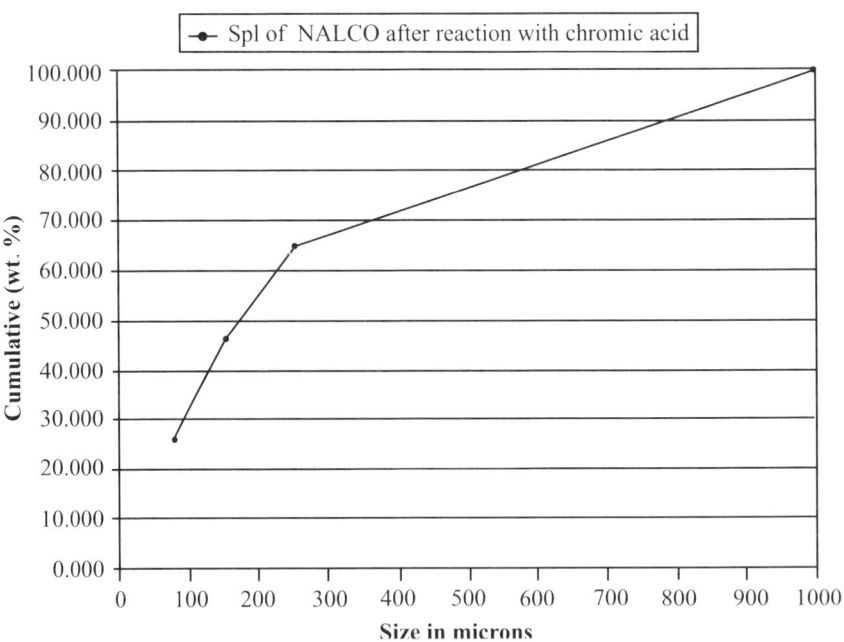

3.12(b) Cumulative graph showing particle size distribution of SPL of NALCO, Angul after reaction with chromic acid.

large value of iron oxide is detrimental towards making any industrial product from red-mud. Moreover, it should be remembered that red-mud is formed from bauxite treatment at high temperature precipitation process.

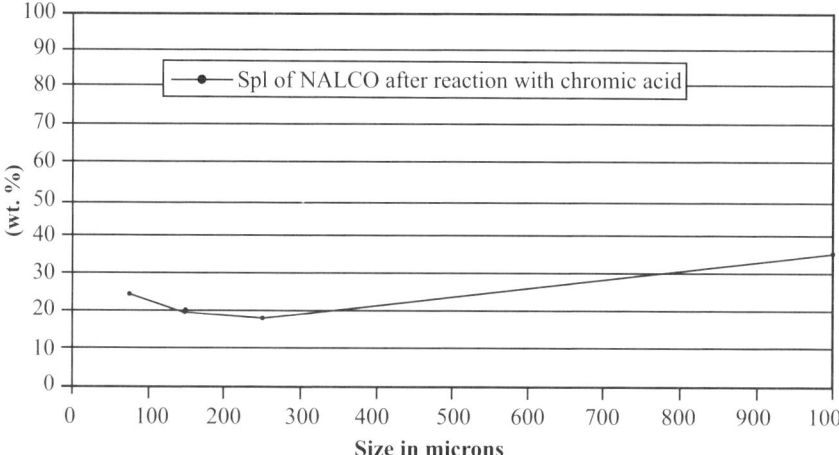

3.12(c) Graph showing particle size distribution of SPL of NALCO, Angul after reaction with chromic acid.

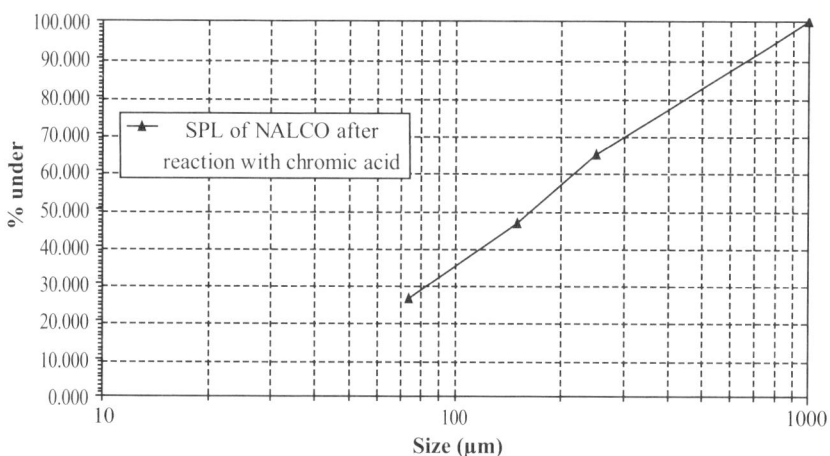

3.12(d) Percentage under size of SPL after reaction.

Accordingly, the particle size of red-mud is fine and the iron is embedded into the crystal lattice of the fine particles. Thus it is difficult to remove all iron from red-mud. No physical method can separate iron from red-mud. Complete removal of iron by chemical means is also difficult and expensive. Separation of iron from rest constituents of red-mud by melting process and recovery of the metal by this process vis-à-vis conventional method of extraction of the metal from hematite ore, is economically not viable.

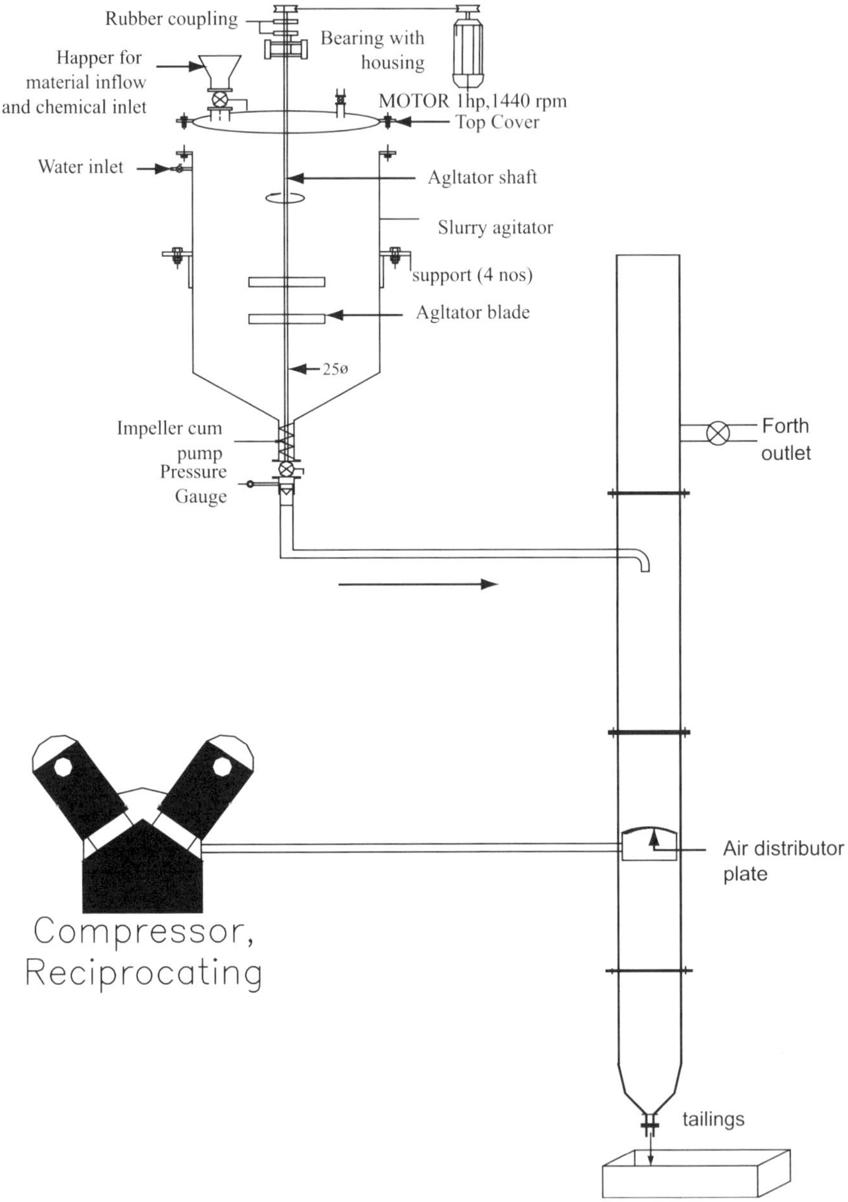

Rubber coupling

Bearing with housing

Happer for material inflow and chemical inlet

MOTOR 1hp,1440 rpm
Top Cover

Water inlet

Agltator shaft

Slurry agitator

support (4 nos)

Agltator blade

←25ø

Impeller cum pump
Pressure Gauge

Forth outlet

Air distributor plate

Compressor, Reciprocating

tailings

3.13 Schematic diagram of flotation column.

Table 3.3 Composition of red-mud generated by different Indian aluminum Plants [8].

Constituents	NALCO	HINDALCO	MALCO	BALCO	INDAL, Muri	INDAL, Balgaur
Al_2O_3	15.00	17.0–22.4	14.0–	18.0–20.0	24.0–26.0	18.0–20.0
Fe_2O_3	62.78	25.6–33.2	18.0–	27.0–29.0	36.0–38.0	40.0–50.0
TiO_2	3.77	15.6–16.5	50.0–	16.0–18.0	16.0–20.0	8.0–10.0
SiO_2	6.55	6.9–8.25	56.0–	6.0–8.0	5.0–6.0	5.0–7.0
Ca O	0.23	5.6–14.6	2.0–4.0	6.0–12.0	0.8–1.0	1.0–3.0
Na_2O	4.88	3.9–5.8	6.0–9.0	4.0–6.0	3.0–3.5	3.0–4.0
Mn	1.10	–	–	–	–	–
P_2O_5	0.67	–	1.0–2.0	–	–	–
V_2O_5	0.38	–	–	–	–	–
Gd_2O_3	0.01	–	–	–	–	–
LOI	9.50	8.5–12.6	–	12.60	–	10.0

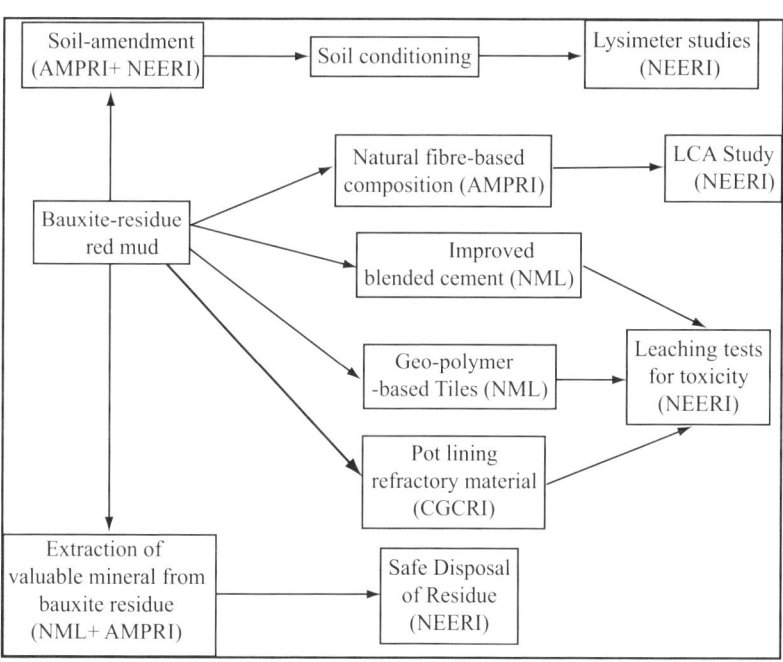

3.14 Total aluminum plant waste utilization as developed by Indian R & D Organization.

Abbreviation: NEERI (National environmental Engineering research Institute, Nagpur).

NML (National Metallurgical Laboratory, Jamsedpur)

CGCRI (Central Glass Ceramic Research Institute, Kolkata)

AMPRI (Advance Materials and Processes Research Institute, Bhopal.)

Table 3.4 Current trend of red-mud disposal by Indian aluminum smelters

Serial No.	Name of the plant	plant capacity (t)	Red-mud(t/t) of alumina	Dumping procedure
1	INDAL, Muri	72,000	1.35-1.45	This refinary adopted the closed cycle (wet slurry) disposal system (CCD). The disposal ponds have not been provided with any liner.
2	INDAL, Belgaum	2,20,000	1.16	The plant switched over to dry disposal mode from wet slurry disposal mode in 1985. The mud after clarification passes through six stage counter current washing and after filtration (65% solids), it is disposed off by dumpers at the pond site. The dry portion of the pond is covered with a 15 cm black cotton soil for promoting green vegetation.
3	HINDALCO, Renukool	3,50,000	1.4	Traditional CCD method of impoundment was used. In late 1979 dry disposal method was implemented. After five stages counter current washing the solid is filtered (70% solids) and disposed off into the pond.
4	BALCO, Korba	2,00,000	1.3	Residue after settling counter currently washed in four stages and filtered. The filtered cake is repulped with the pond returned water and dumped in the pond. Uses modified CCD system of disposal. The dykes of the currently used pond have stone masonry and well protected polythene liner and clay layer.
5	NALCO, Damonjod	8,00,000	1.2	A modified CCD method is used for disposal. Subjected to six stage counter current washing by pond returned water (0.5 g/l Na2O) and condensate from the evaporators. The washed red-mud is repulped and sent to disposal sites. The bottom and sides of the pond are covered by impervious and semi previous clay with base filters.

Most of the smelter plant thus dilutes iron content in red-mud by addition of the other inert materials to red-mud. But then the consequent high throughput of red-mud from aluminum plants does not provide solution to complete disposal by economic means. Dissolution kinetic of iron and aluminum from red mud using acid has also been reported in the literature [206].

High value constituent like titanium dioxide in red-mud justifies its extraction provided concentration of the metal meets economic threshold limit of the compound. Researches are under way to extract titanium from red-mud in certain locations providing economic concentration of titanium dioxide in red mud.

Economic extraction of rare earth elements like Cb, Sc, Zr, Nb, etc. is also limited to very few cases for above reason. Disposal of red-mud as land fill still relies on inverting red-mud with silica, lime etc before use as land fill.

Because of higher iron content, use of red-mud in cement industry is also very limited. Further bound alkali in red-mud possess problem in large scale use of red-mud.

Present authors have attempted to make red oxide pigment from red-mud by extraction of iron from red-mud [207]. The process generates medium quality red-oxide primer and can be sold commercially. Research is underway in author's research laboratory for converting red-mud to electro-porcelain (high voltage tension line insulator) after removing iron from its composition. Vehicle used for making such red oxide paint were linseed oil, varnish, drier and thinner (turpentine oil).

3.3   Treatment of waste emulsion generated by wire-rod mills

As mentioned by earlier wire rod mills housed within smelter plant premises rejects large volume of waste emulsions which because of its toxic oil contents can not be discharged in open drains. Further since above toxic oil contain in the waste emulsion being only around 7% it is imperative that the residual water after breaking the emulsion needs to be recirculated to the plant itself. Present authors developed a process [123] for breaking the emulsion and release of the residual the water conforming to statutory norms for disposal of treated water in the open drain. In this process the waste emulsion was treated with calcium hydroxide in order to coagulate the toxic oil and separate it out from the residual water. Small contaminants in the residual water were finally removed by activated charcoal, pH adjusted to 7 and the water released. Typically for 7% oil content in waste emulsion, application rate of 3 wt%

calcium hydroxide and 2.5 wt% charcoal brought down C.O.D. of treated water within permissible range. Table 3.5 gives an example of such treatment process.

Table 3.5 Typical example of treating waste emulsion by the author [208]

Initial oil content in waste emulsion	Amount of Ca(OH)$_2$ added (in wt %)	Amount of activated charcoal (wt %)	oil in treated water	Aluminum content in the water	C.O.D. value of treated water	pH value of the treated water
6%	3	2.5	25 ppm	nil	60	7

Utilization of wastes (byproducts)

In previous chapter we have dealt in detail procedures for treating major wastes from aluminum smelter plants. Some of this treatment in turn generates byproducts (e.g. carbon powders from spent pot liner and anode butts). In this chapter we will discuss possibilities of making various industrial products from these waste byproduct carbon.

4.1 Byproducts from treatment of spent pot liner and anode-butts

As mentioned in earlier chapter, oxidizing acid treatment of these wastes generates fine carbon powder which may be wholly graphitic in nature or a combination of the same with amorphous carbon powder depending upon the severity of the reaction. Characteristics of these carbon powders have been mentioned in Table 4.1. Close scruting of this table indicates that possibility lies in converting these carbon powders to a number of industrially important products. The same have been investigated by the present author, and are discussed in detail in following sections.

4.1.1 Colloidal graphite and dry lubricants

If the recovered carbon from spent pot liner is predominantly graphitic, it can be used to make colloidal graphite or dry lubricant [209]. Here the advantage is that the derived carbon is already in micron size and thus dose not require elaborate and extensive grinding as required in case of conventional colloidal graphite preparation from natural graphite. However, in most cases it will be helpful to beneficiate these carbon powders before use in order to bring down ash content to desired level. Figure 16 shows set up of a flotation column which can beneficiate these carbon powder. These carbon powders can also be used directly as such for making dry lubricant for application to frictional parts. Dry lubricants are used in high frictional areas generating heat at the frictional points. Since these graphitic powders have high temperature resistant property, they are suitable for such applications. Their lower cost and satisfactory performance gives them an edge over other synthetic dry lubricants. SPL derived carbon powders can also be suspended in temperature resistant fluids, such as silicone oils, to prepare colloidal graphite also used for lubricating frictional parts. Present author in his experiments found loading rate of 15–20 wt% of carbon powder in these oils resulting in viscosity in the range 300–3000 centi-poise and with comparable density and suitable surface charges. Above properties helps these carbon powders to remain in suspended in the oils for a prolonged period and thus avoids formation of bottom sludge on long storage. Silicone oils itself is capable of resisting temperature up to 170°C which dose not set limit to the use of these compounds as the graphite powder is self lubricating and can withstand still higher temperature. The colloidal graphite, unlike hydrocarbon base lubricants, is fire-resistant and has a very long shelf-life.

4.1.2 Production of pyrogenic silica at relatively lower temperature

Pyrogenic silica is termed as micro-amorphous silica, although it is not truly amorphous but consists of regions of ordered or crystals of extremely small size. Common form of micro-amorphous silica has ultimate spherical silica particles of less than 1000 angstrom in diameter, the surface of which consists of anhydrous SiO_2 or SiOH groups. The powders generally occur in coherent condition with a three-dimensional aggregate formed by siloxane bonding. These clusters are disintegrated in solution phase (generally by alkalis like sodium-hydroxide, ammonium solution, etc) to form sols. Absorption of alkalis on silica surface reduces rate of dissolution and thus growth of the clusters retarded. Sols thus formed has great commercial value and a number of these sols (as indicated in Table 4.1 below) for silica particle sizes from 4 mm to 36 nm, finds multiple uses like binders and stiffeners for modifying frictional properties of waxes and fibers, modifying adhesion before surfaces, reinforcing polymers and rubbers, as polishing agent, as viscosity modifier, frosting incandescent bulb glasses, catalyst support, etc.

Table 4.1 Size of silica particles in commercial silica sols listed by manufacturers

Manufacturers	Commercial name of sol	SiO_2 (%)	Particle diameter (nm)	Surface area (m^2/g)
E.I.duPont demour, Wilmington, Del, USA	Ludox	30–40	12–21	230–130
NALCO Chemical CO, Chicago, Il, USA	Nalcoag	15–30	4–15	750–210
NYACOL Inc, Ashland, Mass, USA	Nyacol	30–50	8–20	–

NISSAN Chem Industry, Tokyo, Japan	Snowtex	20–30	10–20	–
Monsanto Ltd, London, England	Styton	25–30	36–18	–

Amorphous silica's with very small particle sizes are prepared by aqueous polymerization process, which involves hydrolysis of gases like $SiCl_4$ and simultaneous polymerization of silica acid in water. Stobber & Fink [210] prepared very uniform spherical silica (from few nm to 900 nm) by hydrolyzing a lower alkali-silicate in an alcohol solution. Pyrogenic silica's on the other hand, can be prepared by vaporizing silica at very high temperature (around 2500°C). If a reducing agent is present simultaneously so to form SiO, the sublimation temperature can be brought down to 1700°C [211]. As the monoxide evaporates in an oxidation atmosphere, the dioxide is condensed in an oxidizing atmosphere; the dioxide is condensed in an extremely finely divided form. A remarkable form of silica "fluffy" form having very low density and appears to flow as water does, was prepared by Jacobson [212] from SiF_4. Present authors prepared [213] this variety of fluffy pyrogenic silica at a much lower temperature (only at 450°C) using the SPL derived carbon powder with commercial grade silicon oil and heating the mixture in a furnace around at 450°C. Fluffy variety of pyrogenic silica thus produced is extremely light (density about 0.025 g/cc), white in appearance, and stable in extreme conditions. The product dissolves instantly in acids like hydrofluoric acid and gets dispersed in water easily by adding a little alkali. Microscopic examination indicates silica particles in a three-dimensional network giving it a fluffy cotton-like appearance. Malvern particle size analysis shows average particle size of 4 micro-meters and a surface area of $1.1575 \ m^2/cc$ ($495 \ m^2/g$ by BET).

4.1.3   Preparation of pencil lead

Pencil leads are essentially baked ceramic rod of clay bonded graphite. The suitability of graphite for pencil industry is judged by the dark streak it leaves on the paper by scratching. Synthetic graphite through has less ash content and very fine particle size, produces very little smear and thus is unsuitable for pencil manufacture. Usually pencil manufactures prefer amorphous graphite with about 90% purity and free from gritty particles. Amorphous graphite gives better smear than the flaky graphite. Indian Standard Institute (ISI) has thus set norm for graphite suitable for pencil industry as amorphous graphite with size -300 BS mesh and ash less than 50%. Micronised amorphous graphite accordingly is very much sought after raw material for the manufacture of high class lead for pencil.

Table 4.2 Characteristics of pencil lead produced from carbon powder derived from spent pot liner [214]

Ingredients	Binder	Firing temperature	Transverse breaking strength
Carbon SPL powder + Bauxite/Kaoline clay	30%	1100°C	200 kg/cm
Carbon SPL powder + local plastic clay	30%	1000°C	225 kg/cm
Carbon SPL powder local plastic clay + phosphoric acid	40%	400°C	250 kg/cm

General procedure followed for making pencil lead from these micronised amorphous graphite involves drying the slurry to required consistency, and form a stiff dough with the mixture, followed by compaction into a solid cylinder to form and extruding under pressure through a die into pencil lead. Wet strands thus received are then dried, packed in saggers and fired at 1100°C . The fired leads are then impregnated with waxes and fats or fatty acids, or both. These latter steps preclude glazing of the point of use. The quality of the finished lead depends on the ratio of clay and graphite. Clay used varies from 20 to 60%. Hard Pencil has about 60% clay. Hardness is a product requirement and not a quality factor such as the uniformly, smoothness or strength of the sharpened point. Indelible leads, on the other hand, are mixture of graphite, methyl-violate, and such as gum or methyl-cellulose, with or without mineral fillers and insoluble soaps.

Attempts were made by the present authors to produce pencil lead from carbon powders derived from spent pot liner [214]. Above procedure for bonding and backing the carbon powder with clays were followed and the results are shown following Table 4.2.

4.1.4 Making dry cell electrodes

Dry cell uses carbon electrodes prepared by mixing various carbon raw materials like amorphous carbon, baked graphitic and electro-graphite and firing them at appropriate temperature. Generally electrode pitch (mesophase pitch) is used to bind these raw materials to impart necessary mechanical strength. Table 4.3 indicates approximate electrical resistance of these raw materials:

Table 4.3 Typical electrical resistance values of raw material used for dry cell electrodes

Material	Typical resistance
Amorphous carbon	48 micro ohm
Carbon graphite mixture	45–50 micro ohm
Baked graphite	25–40 micro ohm
Electro graphite	below 25 micro ohm

Table 4.4 Volume resistivity of carbon powder derived battery cell electrodes

Electrode/binder	Volume resistivity	Mechanical strength
No bonding agent press formed electrode	0.135 ohm cm	1 kg/cm^2
Electrode of carbon powder with polyvinyl alcohols as bonding agent	0.0764 ohm cm	5 kg/cm^2
Electrode of carbon powder bonded with 10% pitch	0.524 ohm cm	9 kg/cm^2
Electrode of carbon powder bonded with 10% zinc	0.045 ohm cm	10 kg/cm^2
Commercially available Eveready dry cell electrode	19.337 ohm cm	10 kg/cm^2

Besides binder pitch, combination of metals (copper, zinc, tin, etc) producing eutectic solution and wetting the graphite surface can also be used provided it does not react with the electrolyte solution in the battery. Present author has tried various type of binders like polyvinyl alcohol, electrode pitch, metal powder etc to form electrode with the derived carbon powder from spent pot liner treatment [215] and the results are shown in Table 4.4 below.

Polyvinyl alcohol added 7 drops of 0.1% solution to 4 g SPL powder to which pitch was first dissolved in an organic solvent and mixed with the powder in required ratio. The product finally heated to 180°C to drive off the solvent. Metal powder after addition to the carbon powder was press formed into a rod and then heated to desire temperature under inert atmosphere. It can be seen from above table that these SPL derived carbon powder are good candidate for making dry cell electrodes as they meet both electrical conductivity and strength requirement for this specific application.

4.1.5 Production of carbon refractories

Carbon refractories are used in metallurgical industries where reducing atmosphere prevail in the operating condition. One such example is the lower hearth (bosch) region of the blast furnace. Stringent property requirement of these carbon bricks include high mechanical strength and abrasion resistance as well as the wetting phenomenon with the molten metal. Present author in his experiments [216] with SPL derived carbon powders found that they are suitable for making carbon bricks. Here the binders used are clays (china clay and ball clay) as well as quartz powder and feldspathic rock. Combination of these ingredients not only imparts necessary mechanical strength after final firing, but also counteract expansion/contraction phenomenon associated with high temperature curing of the green bricks. To obtain a good dense body, it is always advisable to extrude the green body through pug mill under high pressure. The green bricks are first biscuit fired at 600°C and final firing carried out at 1300°C under inert/reducing atmosphere. These bricks easily develop required transverse breaking strength of about 300 kg/cm with above binders to carbon ratio of 1:1. This ratio also ensures necessary wetting property of the bricks to molten metal.

In a typical example of above experiment performed by present author's 6 g bentonite clay was mixed with 4 g of china clay, 6 g of feldspar powder and 4 g of quartz powder all passing through -125 micron mesh. This refractory mix was mixed with 20 g graphite powder derived from SPL. A little distill water was added to the mix and formed into a tight dough which was pressed into a 1 × 1 × 2 in. cardboard tray and kept for 24 hours in a shaded area for curing. The cast brick then was pressed by a weight of about 20 kg to compact the same. The hardened brick then carefully taken out and heated slowly up to 600°C at about 50°C/h incremental rate and hold at 600°C for 6 hours. The cured brick was covered with carbon powder and placed in a high alumina crucible before firing in inert gas atmosphere. The temperature of firing then slowly increased to 130°C and kept at that temperature for about 12 hours. The furnace was then cooled slowly to room temperature and the cast brick taken out of the furnace. This carbon refractory showed transverse breaking strength of 280 kg/cm.

4.1.6 Preparation of silicon-carbide from spent pot liner

Among four types of carbide formed by metals with carbon (e.g. salt like carbides, metallic carbide, and diamond like carbides) silicon carbide falls in the diamond like carbide category. Silicon-carbide because of its inherent hardness, high thermal conductivity and semiconductor behavior finds number of important industrial applications. One of its common use includes forming abrasives (in loose form for lapping, mixed with a vehicle to form abrasive paste or sticks, with binders shaped to form abrasive sheets, disks and belts). Here it may be noted that silicon-carbide is harder yet more brittle than abrasive such as aluminum oxide. Since its grain fractures readily and maintain a sharp cutting action, silicon-carbide abrasives are generally used for grinding hard, low tensile strength materials such as chilled iron, marble and granite, and materials that need sharp cutting action such as fibers, rubber leather or copper. Its extreme hardness is utilized in forming wear resistant products such as brake linings or electrical contacts and in road surfaces. Its low coefficient of thermal expansion and high thermal conductivity as well as chemical inertness is utilized in making high temperature refractories, boiler furnace walls, checker bricks, muffles, kiln furniture's, furnace skid rails, try for zinc purification plants etc. Its excellent electrical properties are utilized in forming electrical heating elements from recrystallized silicon-carbide. Electrical furnaces made from such heating elements, provides service temperatures as high as 1600°C. The semi-conducting properties of silicon-carbide have led to its use as thermistors (temperature sensitive devices) and varistors (voltage sensitive devices). It is also being used in making high temperature furnace thermocouples. Its extraordinary stability finds its use as catalyst carrier nuggets, tower packing and pebbles for fluidized bed reactors. Its other uses include raw material for the production of silicon tetrachloride, in welding rod compositions, as fillers in elastomers, as an additive in high temperature ceramic compositions. Ultra fine silicon-carbide produced in electric arc is used as insulation in cryogenic applications. Its low neutron cross-section and good resistant to radiation damage make it useful in nuclear reactor applications. Other uses of fine silicon carbide powder include preparation of mould coating for batch casting of molten metals.

Properties of silicon carbide depend upon its purity, polycrystalline type and method of formation. Below 1800°C alpha variety is produced. Thus, values reported for commercial polycrystalline silicon carbide should not be interpreted as being representative of single crystal silicon-carbide. Following examples illustrate wide variation in hardness for various grade silicon-carbide.

Table 4.5 Hardness of various grade silicon carbide

Type of silicon carbide product	Knoop hardness
Dense direct-bonded silicon carbide	2740
Sintered alpha silicon carbide	2800
Black single crystal silicon carbide	2839
Green single crystal silicon carbide	2875

Silicon-carbide is commercially produced from silica sand (quartz) powder and petroleum coke (CPC)/anthracite coal in required proportion in an electric furnace. Heat at the core of such furnace reaches as high as 2600°C. A yield of 11.3 ton black silicon carbide is obtained from a furnace charge of 75 ton by this process. For preparing higher purity (green silicon-carbide), fresh and high purity (low aluminum content) raw materials are required. Ultra fine silicon carbide is produced continuously in the electric arc furnace using consumable anodes of silica from rice husk. Hot pressed silicon carbide of high hot strength and density up to 99% of theoretical, may be prepared under pressure (69 MPa, 10000 psi) at 2000–2560°C. Small addition of 1% aluminum assists in compaction and permit use of lower hot pressing temperature. A material made of silicon nitride or silicon-oxynitride bonded to self bonded silicon-carbide has high corrosion resistance and may be used for pump parts, acid spray nozzles and in aluminum reduction cells.

Attempts have been made at Clemson University, South Carolina, USA by Denis A Brosnan [217] to prepare silicon carbide from spent pot liners of aluminum industries. In this attempt the spent pot liner was first crushed and powdered to less than 1 mm particle size. The powdered product was then mixed with silica powder in the molar ratio 3:1 to 5:1 and heated in a closed electrothermic smelting furnace. The final product is silicon carbide as claimed in the process. As mentioned earlier raw spent pot liners contain a large amount of toxic contaminants such as fluoride, cyanide, alkali, etc, which is claimed to breakdown by this process and exit through the fluc gas stream in this process. These toxic fumes emanating from the furnace can then be trapped either by wet or dry scrubbers. However, there is every likelihood that these corrosive gases may attack furnace lining and bring down furnace life drastically. The process also require fine grinding of large hard pieces of spent pot liners which can be achieved by successive use of hammer mill, jaw crusher and Raymond Mills. Thus the raw material preparation consumes lots of electrical powder and is expensive. Contaminants in the final product (silicon-carbide) are reported to have been separated by gravimetric methods.

Above mentioned drawbacks of silicon-carbide formation from spent pot liners have been obviated by the present author in a new technique [415] whereby the spent pot liners are decontaminated by reacting with strong oxidizing acids as mentioned earlier and then subjected to high temperature reaction with silica powder in an argon plasma furnace. The raw material being cleaned before high temperature reaction, no toxic gas emanates by this process, nor the furnace lining affected by the emanating gases. As mentioned earlier these kind of oxidizing reactions (e.g. with chromic or perchloric acid) dose not need very fine powder preparation from the spent pot liner and a size of half inch to dust is sufficient to carry out the reaction in a short time. The reaction product generates by itself very fine carbon powder (average particle size 20 micron) and thus the reaction is easy as well as homogeneous at high temperature. Moreover an argon plasma furnace has a very compact hearth configuration and can process very large amount of powder in a small fraction of time in-flight condition. The recovered carbon from chemical treatment of spent pot liner is mixed with -200 BS mesh size commercially available silica powder in the molar ratio 3:1 to 5:1 and being charged into the furnace continuously using argon as a career gas. The reaction time is few minutes and the product accumulates at the bottom in a graphite crucible. The product is collected from the bottom of the furnace and characterized by XRD analysis.

In such an argon furnace the plasma is created by passing an electrical current through argon gas. The pot-type DC extended arc plasma reactor used in this case has its extended arc plasma created by ionizing argon gas under its arc zone. The gas itself is introduced continuously into the arc zone through an axial hole of one of the electrodes. Once the arc is stabilized, it is then possible to extend the arc length maintaining the same power level. The pot type extended arc plasma reactor used in the present process operators at a 50 KW DC rating. Figure 4.1 shows schematically the internal construction of the plasma furnace. As can be seen in above picture, two graphite electrodes are arranged in the vertical configuration. The graphite crucible is used as the hearth of the hearth of the anode in the form of graphite crucible which also serves as the hearth of the reactor and a graphite electrode on top of it as an axial hole for continuously passing plasma forming gas into it. During operation, this bottom electrode and the crucible is kept fixed and the formation as well as stabilization of the extended arc plasma is carried out by moving the top electrode through a rack and pinion arrangement.

The hearth is thermally insulated by the bubble alumina contained in a mild steel casing. The dimension of the graphite hearth is typically 100 mm outer diameter, 70 mm inner diameter, 200 mm long with a depth of 170 mm. The top graphite electrode is 25 mm diameter, with 5 mm axial hole. The power of the reactor is supplied by a 50 KW DC power supply. Further a powder feeder is attached with the arrangement of the top electrode (i.e. cathode) to feed the raw material mix in the form of fine powder into the hearth. Too much fine powder may also lead to carry over by the gas current.

During operation, water was passed through the surrounding cooling tube and the prepared mixture of treated spent pot liner carbon and silica

Schematic diagram of a plasma reactor

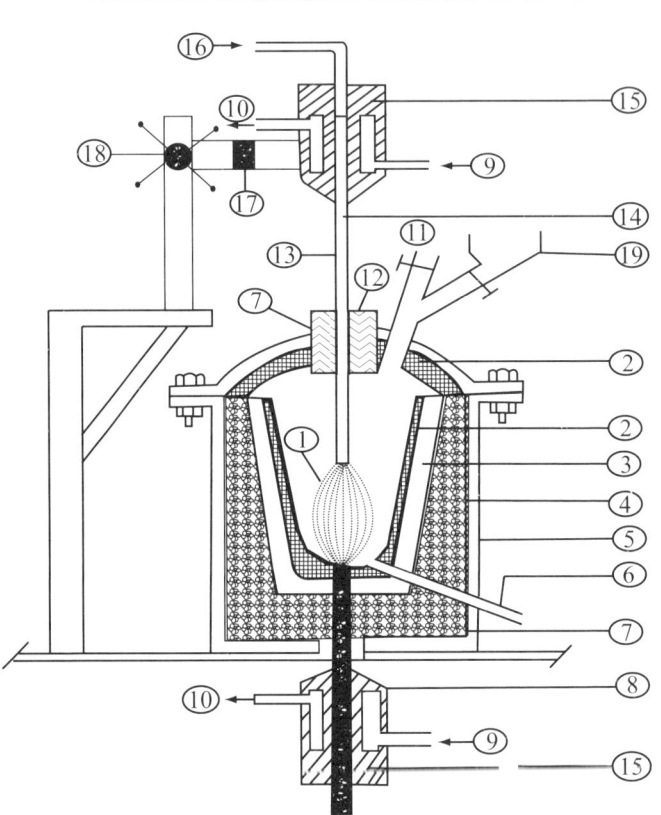

1. Plasma 2. Magnesia Coating 3. Graphite Crucible 4. Bubble Alumina 5. M.S. Casing 6. Tap Hole 7. Alumina Bush 8. Bottom Graphite Electrode 9. Water Inlet 10. Water Outlet 11. Outlet for Exhaust Gases 12. Graphite sleeve 13. Top Graphite Electrode 14. Axial Hole 15. Copper Block 16. Plasma Forming Gas 17.Electrical Insulation 18. Rack and Pinion Mechanism 19. Hopper

4.1 Schematic diagram of indigenously developed extended arc plasma reactor.

was introduced from top of the furnace by carrying gas such as argon. The high temperature condition and additionally a holding time of 2–3 minutes complete the reaction. Figure 4.2 shows the XRD analysis of the product. As is evident from the XRD analysis more than 80% of the mixture is converted in to silicon carbide in such a short span of time. The product is basically in granular form.

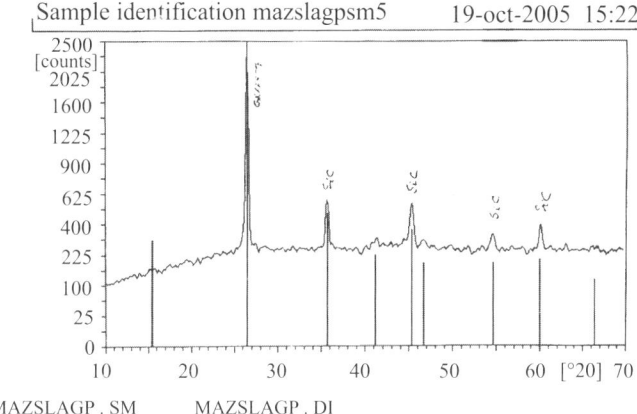

Sample identification mazslagpsm5   19-oct-2005 15:22

4.2 XRD of silicon carbide formation from SPL.

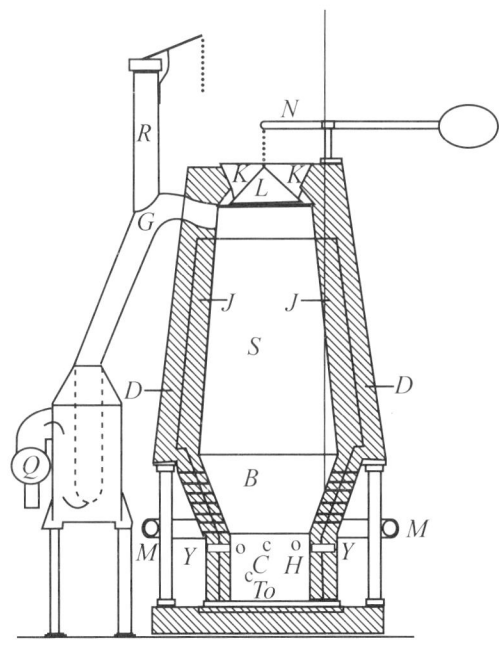

4.3 Diagram showing position of tap hole (T) in conventional blast furnace. S = shaft, B = boshes, Y = tuyere hole, M = hot blast main, H = hearth, G = gas main, J = shaft lining

4.1.7   Preparation of blast furnace tap hole and runner/ trough mass

Molten iron produced in a blast furnace is tapped through a hole situated at the bottom of the hearth (see Fig-4.3). During tapping, an iron bar or an electrical drill is used to dig out the plug used for stopping molten metal from coming out.

Table 4.6 Commercial blast furnace tap hole mass producers

Name of the company Location / address
Sarvesh Refractory Rourkela, Orissa
Raj Ceramics Ranchi, Jharkhand
Janoo Refractory Thane, Mumbai
Unotherm Refractory Airily, Mumbai
Maithan Ceramic Dhanbad, Bihar
Bharat Refractory, New Delhi
Mayerton Refractory Thane, Mumbai
Saint-Gobain Ceramic, UK
Nippon Crucibles, Japan
Riverside Refractory, Poland
Hoor Mehr Sepahan Co., France
Sino-Global Sourcing limited, Hong Kong

When the molten metal has run out through this hole another plug is formed to stop the hole by means of a mud gun. Thus this material is used for closing and opening of the blast furnace tap hole. Tap hole mass mentioned above are generally 1.2 to 2.0 m in length. The liquid metal thus released through tap hole flows through a guided path called "Runner Mass". Tap hole mass being exposed to hot turbulent molten metal, needs to be abrasion and corrosion resistant beside being able to withstand metalostatic pressure above it. While it needs to be strong enough after carbonization, it should also be soft enough to collapse during drilling or pocking by iron rod for tapping the hot metal. Some of the tap hole mass producers in India and abroad are listed in Table 4.6 below. Present day specific consumption rate of tap hole mass is 0.5–0.8 kg/ton hot metal produced [218].

Composition of tap hole mass

Since the tap hole mass remains exposed to molten iron metal at high temperature (1500–1600°C) its major components are refractories like silicon carbide, silicon nitrides, mulite, alumina, silica, etc. Other materials which are also considered for such applications are bauxite, manganese, ferro silicon, graphite, crystalline carbon, etc. Carbon is an important constituent of tap hole mass composition as it provides suitable refractoriness as well as reducing surface to hot metal. Above components of tap hole mass are bound together by binders such as pitch or synthetic resin. These binders cause wetting of the bricks around the tap hole and thus adhere to the surrounding wall of the tap hole. As the mass dry at furnace temperature, they gain sufficient sintering strength as to resist the ensuing metallostatic pressure in the blast furnace hearth, but at the end of carbonization remains soft enough to allow for smooth drilling. However tar although the chief binder it takes longer

Table 4.7 Fusion range of some refractory bricks

Refractory material Fusion range
Fused alumina 1750–2050°C
Bauxite 1732–1850°C
Carborundum (recrystalized) 2200°C (Decompose)
Native clay(USA) 1590–1700°C
Kaolin clay 1645–1785°C
Magnetite 2200°C
Chrome 1950–2200°C
High alumina clay 1802–1675°C
Graphite clay 1605–1675°C
Mullite 1680–1800°C

time to sinter and gives rise to smoke and dust during carbonization process. Obviously, the tap hole mass should have sufficient permeability to allow this evolved gases to escape outside, failing which the mass may collapse. Porosity of the sintered tap hole composition is controlled by careful selection of aggregate particle size distribution which allows necessary gap and packing density. In commercial practice these size distribution have been optimized in the size range 0.4–3 mm and volatile matter content around 12% with consequent bulk density of above 1.8 g/cc to achieve sufficient strength during sintering. Cold crushing strength of the mass is controlled around 22.5 kg/cm at 900°C and 32.2 kg/cm at 1200°C. Instead of pitch if we use other binders such as 'phenolic resin' no smoke/dust is produced but the mass becomes too hard after carbonization. Accordingly, suitable combination of both the binders is used in suitable proportion. Phenolic resins works with ketons, quinoline resin, hexamethylketon and p-formaldehyde and are suitable for such applications. It is also imperative that the viscosity of the generated mass needs to be controlled within a certain range in order to be able to use it in mud-gun. For this reason, viscosity modifiers such as ethylene glycol, diethyl glycol and propyl glycol are also being added in the commercial formulations. Thermosetting resin stable up to 200°C has been used along with coal tar pitch successfully for such formulations [219]. Softening range of the aggregate is generally maintained with a PCE value of 1400°C. As we know PCE value reflects time-temperature behavior of refractory materials compared against standard refractory cones. Fusion ranges of some standard refractory materials are shown in Table 4.7 below.

Fusion or softening temperature [220] is another measure or determining suitably of a refractory under particular service temperature yielding particular mechanical strength. Fusion of refractory materials is not clear cut, but a more or less gradual transition from solid to liquid. The amount of liquid that can be tolerated by a refractory and still leave it in serviceable condition is largely governed by the viscosity of the liquid and type of crystallization present in the solid phase. For example, fireclay refractory may develop liquid and actually start to soften as low as 1880°F (982°C) but due to the high viscosity of the liquids their limiting service temperature may be several 100°C higher. An arbitrary procedure, therefore, evolved is time temperature softening characteristics, commonly called PCE test (Pyrometric Cone-equivalent test), where softening behavior of the refractory is compared against certain standard cones whose time-temperature characteristics are known and numbered. The P.C.E. number to the test refractory is allotted against the standard cone number whose tip touches the supporting plaque. Because of composition of standard cones and temperature range covered, PCE test seems most applicable to alumina–silica refractories. The maximum service temperature of fireclay brick are considerably below PCE temperature, but low PCE ladle brick are successfully used at temperature 300° F above their PCE temperature since exposure is seldom for more than 1 hour at a time. Thus PCE test is not satisfactory for silica and basic brick, because silica brick approach a sharp melting similar to pure compound than refractories. Its crystalline structure also remains rigid to within few degrees of complete fusion. Reversible volume changes of refractories with temperature is called dilation. Silica bricks show a permanent volume change expansion of 13% on firing. Alkalis with fireclay from alkali aluminosilicate with an accompanying permanent expansion of 12% with surface pooling. Another example is iron oxide with basic refractory.

Accordingly, commercial tap hole mass contains alumina–silica refractory around 65–85% rest being carbonious material. Commercial formulation uses pitch anywhere between 1% and 30%. Tap hole mass are generally used for small and medium blast furnace and they contain no tar. Formulations in medium size furnaces (capacity) around 6000 ton/day mainly use alumina and some silicon carbide as aggregates and resins as binder. Formulations for small size furnaces (capacity ranging 3000–4000 ton/day) uses mainly carbon and silicon as aggregates.

Typical compositions of some commercially available tap hole mass are

(a)　Saint-Gobain Ceramics, UK
　　　Silicon Carbide + Carbon = 40%
　　　Silica + Crude silicon = 15%
　　　Alumina (Bauxite) = 29%
　　　Binder = 16%

(b)　Nippon Crucibles
　　　Refractory aggregates = 60–85g
　　　Graphite + Silicon Carbide =5–30 g
　　　Clay = 5–15 g
　　　Binder = 15–25 g
　　　Organic fiber = 0.01–0.75 g

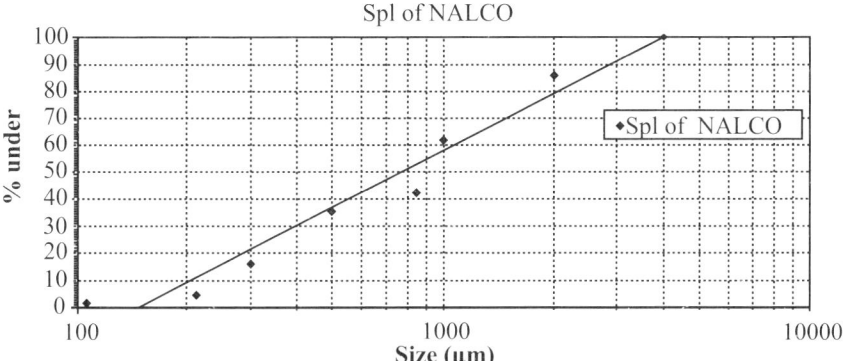

4.4 Average particle size (d_{50}) of crushed SPL (d_{50} = 780μm).

(c) Riverside refractories (Plant of CHZONOW)
     Silica = 65% (Grain size = 0–3 mm)
     Density = 2.2
     Total Carbon = 20%
     Alumina = 10%
     Loss on ignition = 10–20 %

(d) Cherepovetsk steel plant (Russia) (Vanchikov, 1965)
     Clay = 16.7%
     Ground Coke = 50%
     Coal tar pitch = 16.7%
     Grog powder = 16.6%

(e) ACC Refractories (Mumbai)
     Alumina% = 7.5%
     Silica% = 90%
     Iron oxide% = 0.8%
     Titanium oxide% = 0.5%

Use of SPL in tap hole mass formulation

The carbon powder recovered after treating SPL with strong oxidizing acids (mentioned earlier) has following composition.

(a) Ash composition = 10–25 % (for hand-picked SPL)
(b) Alumina = 20–52%
(c) Silica = 42–72%
(d) LOI = 1–2%

   Particle size distribution of the carbon powder is shown in Fig 4.4 below:
   Custom-made tap hole mass composition can be developed using statistical DOX programme [221]. Increase in production rate in molten iron and

Table 4.8 Control parameters for production of tap hole mass [222]

Parameter Specification
Grain size 0–3 mm
Volatile matter 12%
Workability at 100°C 73%
Alumina 10% (maximum)
Fe_2O_3 1% (maximum)
Bulk Density 1.5 g/cc (minimum)
Cold crushing strength 20 kg/cm² (minimum)
PLC% after cocking at +1.00 (minimum)
800°C for 2 hours
Ignition loss 6–20%

increase in tapping temperature 1450–15500°C has been observed in last 20 years. Higher temperature with prolong tapping time and the requirement of constant flow, safety and reduction in specific refractory cost have lead to the development of new tap hole masses. This new tap hole masses contain increased amount of alumina and silicon carbide, up to 50% and 15–20%, respectively. Simultaneously, silica decreased from about 60% down to less then 10%, and about 15% carbon. Specifications of other parameters are shown in Table 4.8 below.

Synthetic alumina has been successfully used along with silicon carbide in various industries [223]. Organic fibers (3–5 mm in length and 5–200 µm in diameter) such as polyester fiber, polyvinyl alcohol fibers, acrylic fibers, polyvinyl chloride fiber, acetate fiber, rayon fiber, polyamide fiber, poly ethylene fiber, polypropylene fiber, polyurethane fiber and poly vinylidene chloride fiber have also been used successfully used [224].Organic fiber generally has MP 250°C or lower. The fiber increases lubricity during extraction when dispersed uniformly. Here at IMMT (BBSR) work is progressing towards making a formulation for tap hole mass where up to 15% recovered SPL carbon powder can be used. Results of such experiments are shown in Table 4.9 below.

Trough or Runner masses on the other hand are developed based on a different set of requirements. Points to note for development of such compound include

(a)   Erosion resistance capability against molten iron flow.
(b)   Corrosion resistant against slag.
(c)   Non-wet ability to iron and slag.
(d)   Low porosity to resist metal penetration.
(e)   Thermal stability and shock resistance.
(f)   Structure integrity at molten iron temperature.

Further, properties of tap hole mass and runner mass are tailor made to suite the individual requirement of the blast furnace based on its capacity from mini-blast furnace of capacity 175 m^3 to large blast furnace capacity 4000 m^3 as mentioned above.

Table 4.10 shows physical properties demanded by a runner/trough mass to meet above requirements.

Thermal properties (PLC %) after coking at 800°C for 2 hour + 2 (+1 min. specified) with tar. Table 4.11 below shows typical chemical composition found with commercially available trough masses.

Table 4.10 Typical physical properties of runner mass

Maximum grain size	3 mm
Bulk Density	2-2.2 g/cc
C.C.S at 900°C	60 kg/cm^2
C.C.S at 1400°C	80 kg/cm^2
Apparent density (after coking at 800°C per hour)	1.8–1.98 g/cc
Permanent linear shrinkage at 1350°C per hour	0.8 (at 900°C) + 0.10 (at 1400°C)

Table 4.11 Typical chemical composition of commercially available trough masses

Volatile matter	8%
Alumina	60%
Silicon carbide	18%
Fixed carbon	14%

Table 4.12 Results obtained for the blast furnace tap hole mass composition with 5% SPL carbon powder by the authors (Patent to be filed by the author soon).

Property	Commercial tap hole mass	Tap hole mass produced from 5% SPL carbon powder
Bulk density	1.5–1.8 g/cc	1.6 g/cc
Fe$_2$O$_3$	1% maximum	Less than 0.5%
Cold crushing strength	20–40 kg/cm^2	30 kg/cm^2
PLC % after coking at 800°C for 2 hours	+ 1.00 minimum	-do-
Alumina content	10% maximum	Within limit
Silica content	10% maximum	-do-

4.1.8 Preparation of aluminum plant anodes

As mentioned earlier Aluminum metal is extracted from its raw material bauxite which is converted to cryolite, by molten salt electrolysis. This procedure, known as Hall-Heroult Process, employs carbon cathode (called pot liner) and an expendable anode which besides conveying current generates heat through resistive load and keeps the electrolyte fluid.

At the end of their use, the left over part of the anode is called "butts" and are crushed and recycled to produce new anodes. However, while recycling the butts the built-up of sodium concentration is closely checked and controlled in order to avoid harmful effect of excess sodium in newly formed anodes. Good quality anodes should be oxidation resistant in gaseous environment in order to minimize oxidation on the expose surface ($C + O_2 \rightarrow 2CO$). Anode butts also contains considerable amount of fluoride.

While preparing recipe of anode making, spent anode butts are mixed keeping in view that above range of properties required for prebaked anodes, so that the anode works satisfactorily in service, external addition of carbon percent are being controlled. For this reason most of the smelter plants have restricted use of butt maximum to the extent of 5% in regular recipe.

In present work replacement of carbon in regular anode recipe was affected by using carbon powder obtained after decontaminating spent pot liner; with the hope that success of such an attempt will open up an avenue for captive utilization of spent pot liner [225]. Earlier present authors invented [1] a process by which all contaminants (e.g. fluoride, cyanide, aluminum etc.) can be extracted out of spent pot liner and byproduct carbon obtained as as fine powder by the process.

Table 4.13 Properties of carbon powder obtained after decontamination of spent pot liner with strong oxidizing acid

Property	Value
Fixed carbon	88%
Ash content	7–16 % (hand picked SPL) 2% (after column floatation)
Specific resistivity	0.135 ohm cm
F$^-$ concentration	140–170 ppm
Na concentration	0.5–0.7 %
CN – concentration	0.1–0.2 ppm
Al concentration	10–15 ppm

Table 4.14 Properties of pure natural graphite

Property	Value
Crystal density	2.266 g/cc
Bulk density	1.3–1.95 g/cc
Thermal conductivity	2000 (in plane) 10 (across plane)
Resistivity (Ω m)	4×10^{-4} 6000×10^{-4}
Spacing between layers	0.3354

Table 4.15 Results of blending 5% spent pot liner derived carbon to regular anode recipe [225]

Property	Values (Range)
Green anode (unbaked) density	1.69 g/cc
Electrical resistivity	47 (μ Ω m)
Baked anode density	1.53 (g/cc)
Thermal conductivity (W/M°K)	7.8 (W/M°K)
Ash %	0.3%
Sodium content	100 ppm
Fluoride	5 ppm
Compressive strength	40 MPa
Baking loss	9.5%

It may be interesting here to compare above properties of spent pot liner derived powder with that of pure natural graphite. This comparison is shown in Table 4.14 below.

Keeping in view of above properties of spent pot liner derived carbon and standard values that of commercial prebaked anode, attempt was made to blend 5% spent pot liner in the regular recipe of anode carbon. The green anode thus made was first dried at 50°C and then baked at 1100°C for several hours. Prebaked anode thus obtained was tested for its properties and the results are shown in Table 4.15 below.

As can be seen above, blending spent pot liner derived carbon powder with regular anode carbon recipe gives comparable density, strength, and electrical conductivity values of commercial prebaked anodes. Thermal conductivity and L_c values are slightly higher in spent pot liner carbon mixed recipe and is probably due to more crystalline carbon in spent pot liner derived carbon powder than baked CPC. However, these two higher values are not detrimental towards performance of the anode, rather may improve anode performance in real trial and thus will prove success of addition of spent pot liner carbon powder in anode carbon recipe.

4.1.9   Absorber for heavy metals, dyes and oils

The recovered carbon powder from SPL was studied for its potentiality in absorbing wastes like oil, heavy metals and dyes. Amorphous carbon of such origin are unique and versatile absorbent as they contain extended surface area and high degree of surface reactivity. In case of graphite, which has a highly ordered structure, the absorption is determined mainly by this dispersion component due to London Forces. However, activated carbon have a random ordering of the aromatic sheets which cause a variation in the arrangement of electron clouds in the carbon skeleton and result in the creation of unpaired electrons and incompletely unsaturated vacancies which influence its absorption behavior. Presence of oxygen on carbon matrix, on the other hand, imparts important property of adsorbing polar molecules either in liquid or gas phase. Although the precise nature of these oxygen functional groups (like carboxylic, ketonic, aldehydic, phenolic etc.) are not known exactly, they are present mainly at the edges and corners of the giant aromatic sheets and constitute the main adsorbing surface [226]. In fact, oxygen atoms located at the surface and edges of carbon crystallites have residual valences which make these atoms act as active site. Adsorption of nonpolar vapors on oxides of active carbon involves purely dispersion interaction. Since the above recovered carbon from SPL constitutes both graphitic and amorphous carbon crystallites, it is expected to be equally effective in both polar and nonpolar chemicals [227].

The effluents from textile, leather, food processing, dyeing, cosmetic and paper industries are main source of dye pollution. Many dyes and their breakdown products may be toxic to living other organisms. Dyes absorbs and reflect sunlight entering water and so can interfere with the growth of bacteria and hinder photosynthesis in aquatic plants. Therefore, decolorization of dyes is an important aspect of waste water treatment [228, 229].

Effluents from some industrial plants like wire-rod mills contain a large amount of waste oil. These oils are sometime taken care by lime, but even then some oil are still left in the waste water which is difficult to remove. Such non-polar substances can also be taken away by activated carbon which is cheaper and simpler than conventional chemical cleaning procedure.

Another common pollutant from industries is the heavy metals in water. Because of low reactivity of some heavy metal ions to oxygen, simpler aeration is not effective technique below pH 9. However, attempts have been made to successfully adsorb these metals by activated carbon. In these trials, effect of adsorbent dose and contact time is commensurate with theoretically

Table 4.16 Summary of oxidation reaction and recovered carbon quality (Fluoride content in starting SPL = 7%)

Oxidizing acids	Fluoride in carbon (%)	Average particle size	Yield (%)
Chromic acid	0.014	20 micro-meter	30%
Perchloric acid	0.060	20 micro-meter	51%
Nitric acid	1.800	35 micro-meter	67%
Alkaline KMnO$_4$	4.700	60 micro-meter	90%

predicted value [230–232]. Adsorption of Hg(II) ions by these methods does not depend upon the surface area alone but appears to be influenced by the presence of oxygen groups on the surface introduced during oxidation surface modification. Adsorption by such activated carbon has been found to increase with oxidation, while degassing (heating between 400 and 950°C) causes decreasing adsorption. Lead and cadmium are other two heavy metal pollutants released by industries, especially by manufacturers of electronic printed circuit boards. Activated carbon has been found to be affective in absorbing these toxic heavy metals as well.

Experiments were designed by present author to measure efficacy of absorbing oil, heavy metals and dyes by carbon powder derived by above oxidation treatment of spent pot liners from aluminum industries. Table 4.16 lists the physical properties of SPL derived carbon powder and the oxidizing agents and used in absorption experiments.

Oxidation of graphitic carbon is known to generate mellitic acid, which as mentioned earlier can be converted to free flowing carbon powder by thermal shock treatment.

For determination of dye adsorption, methylene blue (structure shown in Fig.-4.5) was selected. A solution of 100 ppm of methylene blue in water was prepared by standard method. The thermally treated powder, as mentioned above, was named as SPL2 and non-thermally treated carbon powder was designated as SPL1. Twenty ml of the standard methylene blue dye solution was taken in a beaker and varying amount of SPL 1 and SPL 2 was added to

4.5 Structure of methylene blue.

Table 4.17 Adsorption of methylene-blue dye by SPL derived carbon powder (UV-visible spectroscopy)

Sample	Amount of carbon power added (g)	Amount of dye solution (ml)	Transmission at 635 nm
SPL-1	0.2	20	3.913
SPL-1	0.4	20	2.635
SPL-1	0.6	20	0.144
SPL-1	0.8	20	0.088
SPL-1	1.0	20	0.032
SPL-2	0.2	20	2.150
SPL-2	0.4	20	1.230
SPL-2	0.6	20	0.020
SPL-2	0.8	20	0.050
SPL-2	1.0	20	0.050

it. After keeping the solution for about 15 minutes, the samples were analyzed by UV-visible spectrophotometer (Shimadzu D-1700).

Result of the absorption characteristic at 635 nm is shown in Table 4.17 below.

For measuring oil absorption by SPL1 and SPL2, a pre-treated waste emulsion (containing about 63 ppm oil in emulsion form) was taken as starting oil solution and varying amount of SPL1 and SPL2 was added to 20 ml waste emulsion solution with stirring for about 20 minutes. The filtered solution was estimated for oil content by luminescence spectrophotometer. Results are solution in Table 4.18 below.

Heavy metals absorption studied were done by making solution of heavy metal salts like $Cd(NO_3)_2$, $HgCl_2$ and $Pb(NO_3)$ and Na_3AsO_4 in water to 100 ppm. Varied amount of SPL-1 and SPL-2 was added to aliquot part of the solution and then the solution was filtered and metal ion concentration determined by Atomic Absorption Spectroscopy. Results are shown in Table 4.18 below.

Table 4.18 Absorption oil by SPL1 and SPL2 (Luminescence Spectroscopy)

Sample No.	Original Oil concentration	Amount of carbon powder used	Absorption time	Oil concentration after absorption
Blank	63 μg/l	–	–	–
SPL-1	63 μg/l	4% wt/wt	15 min	33 μg/l
SPL-2	63 μg/l	4% wt/wt	15 min	35 μg/l

Table 4.19 Absorption of heavy metals by SPL-derived carbon powder (Atomic Absorption Spectroscopic study)

Sample	Metal ion (Concentration)	SPL carbon added	Adsorption time	Concentration of metal after adsorption (ppm)
SPL1	Na_3AsO_4 (100 ppm)	1 g/20 ml soln	8 hours	35.75
SPL2	- do-	-do-	-do-	43.84
SPL1	$Pb(NO_3)_2$ (100 ppm)	1 g/20 ml soln	8 hours	7.05
SPL2	-do-	-do-	-do-	8.40
SPL1	$Cd(NO_3)_2$ (100 ppm)	1 g/20 ml soln	8 hours	20.00
SPL2	-do-	-do-	-do-	22.00
SPL1	$HgCl_2$ (100 ppm)	1 g/20 ml soln	8 hours	50.00
SPL2	$HgCl_2$ (100 ppm)	-do-	-do-	52.00

Oil absorption values are shown in Table 4.18. As the result indicates almost 50% oil is being absorbed by the carbon powder. SPL1 shows slightly more oil absorption power than SPL2 on comparative basis. These could be due to presence of carboxylic group in SPL1 which participate in coupling with oil in emulsion phase.

Absorption data on dye (methylene blue) by SPL1and SPL2 is shown in Table 4.18 above. The results are also represented graphically in Figure- 4.6 and Figure-4.7. As the results indicate at the application rate of about 0.7 SPL carbon powder for 20 ml dye solution, absorption of dye by the carbon powder (both SPL1 and SPL2) is almost complete. The result on other hand manifests strong adsorption power of the SPL derived carbon powder towards organic dyes.

Table 4.19 shows absorption of heavy metal ions by SPL derived carbon powder.

In all cases absorption varies from 50 to 90%. Again on a comparative basis absorption of metal ions by SPL1 is more than SPL2. Most probably this is due to occurrence of carboxylic groups in non-thermally treated SPL1 than in thermally treated SPL2. Polar groups in SPL1 attract heavy metal ions in solution phase as ions. Earlier workers have conclusively proven that acidic groups on surface of carbon powders plays vital role in determining adsorption properties with polar or ionic groups in solution phase [233–235].

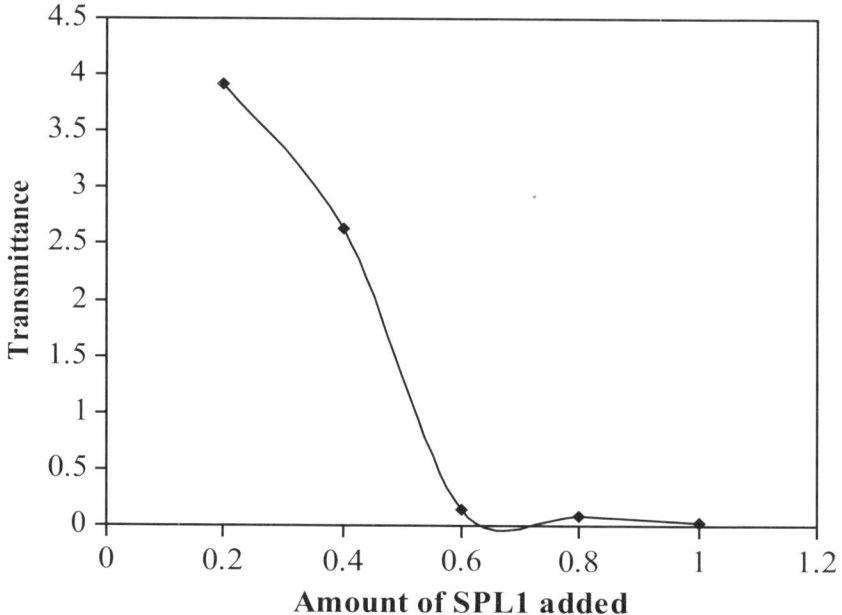

4.6. Adsorption of methylene blue dyes by SPL1 derived carbon.

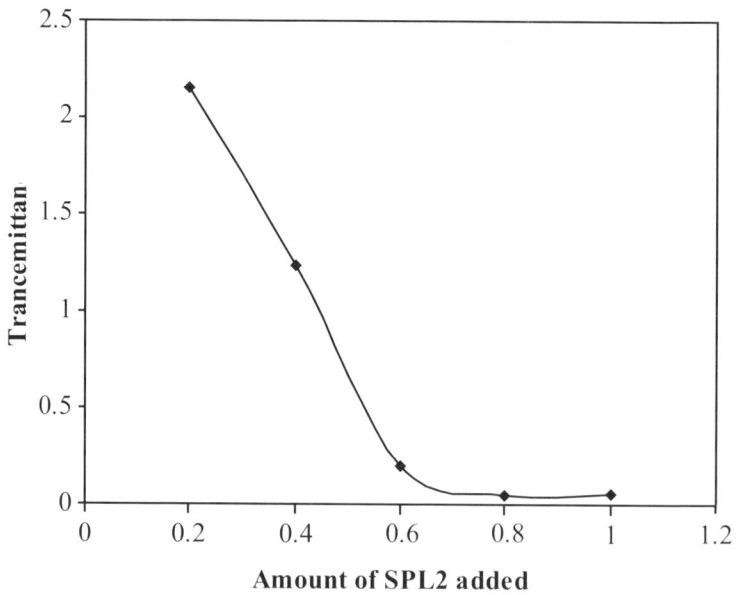

4.7. Adsorption of methylene blue dyes by SPL2 derived carbon.

The study thus indicates that carbon powder derived from spent pot liners, a waste material of aluminum smelter plant, can be utilized to absorb dyes, oils and heavy metals in drain water of various industries thereby can be considered a potential product for fighting industrial pollution.

4.1.10 Preparation of mould coating for batch casting of iron and steel

Byproduct carbon obtain from treating SPL can be used to manufacture mould coating for batch casting of ferrous metals in permanent moulds [200]. The mould coating was prepared by mixing the SPL derived by product carbon powder with china clay, glass powder, sodium dihydrogen phosphate etc. In this preparation the SPL-derived sieved through -200 BS mesh size and above ingredients were also prepared to comparable mesh size. The mould coating is prepared at the application side (casting bay) by mixing with aqueous solution of organic binder such as gum acacia. After Appling the coating in side the mould by brush, it is dried in the temperature range 60–70°C, and the mould is ready for batch casting. Trials by the present author at a renowned steel production unit in India indicated 100% striping value and excellent surface finish. Stripping value indicates the ease of taking out the cast from the mould while surface finish is generally determined by visual inspection.

4.1.11 Recovery of fluorine as aluminum fluoride

Aluminum fluoride is used directly into electrolytic cell for control of fluoride concentration, thereby increasing electrolytes conductivity and eutectic temperature. Aluminum fluoride is a high value product and numerous attempts have been made for recovery of fluorine from spent pot liner as aluminum fluoride for return to the smelter plant.

Aluminum fluoride (AlF_3) is dimeric in liquid as well as gaseous state and occurs as trigonal molecule of D_{3h} Symmetry. The Al-F bond length is 163 pm. Its other properties are shown in Table 4.20 below.

Table 4.20 Properties of aluminum fluoride

Molecular formula	AlF3
Molar mass	83.977 g.mol^{-1}
Appearance	White, odorless crystals
Density	2.91 g/cc (solid)
Melting point	1290°C
Solubility in water	Insoluble

It may be noted here that active alumina is capable of reducing fluoride content in water. Optimum fluoride concentration in drinking water is allowed up to 1.5 mg/l in Estonia and European Union [236]; whereas a study in USA showed optimum fluoride concentration in temperate climate as 1–1.2 mg/l [237]. Other materials used as filtering medium to remove fluoride in drinking water are calcite and limestone. Ion exchange method of fluoride removal from water is effective only if fluoride content in water is less than 10 mg/l [238]. For this purpose fluoride determination in various matrices has been elaborated by Campbell [239]. After dissolution of fluoride, fluoride has been determined by ion selective electrode, photometric method, ion chromatography, atomic and molecular spectroscopy.

Fluoride in spent pot liner has been separated basically by two processes. In the first method fluoride in SPL is converted to HF by acid treatment and then distilled off. In second case fluoride has been separated by wet chemical method. Fluoride gas emanated from aluminum electrolytic cell is absorbed by dry scrubbing. By using smelter grade alumina (primary alumina) as scrubbing medium (adsorbent), fluoride is captured as AlF_3 and returned to the plant. Wet chemical methods include dissolution of fluorine-compounds in both basic and acidic solution. The dissolved fluorine compounds then recovered either as AlF_3 or cryolite (Na_3AlF_6). In pyrohydrolysis method of SPL treatment, SPL is paralyzed in conjunction with introduction of water to produce an off-gas containing fluoride material as exemplified by U.S. Patent 4,113,832. Such pyrohydrolysis technique may also be carried out in a fluidized bed reactor (U.S. Patent 4,158,701 and 4,160,808). These processes unfortunately still tend to produce large volume of waste with high non-leachable fluoride and safe only to use as a landfill.

Wet method of fluoride recovery while completely removes leachable fluorides and to a large extent non-leachable fluoride, ultimately produces a solid end product suitable for brick or cement industry as well as suitable as fuel. These processes are generally carried out in a acid digester where some fluorine escape as HF along with the cyanides in this phase can easily broken down by heating and the fluorine compounds recovered. Caustic treatment of SPL on the other hand recovers fluoride as sodium fluoride.

Readers interested in more details for AlF_3 manufacture from spent pot liners can refer references [240–280] for further details of the processes.

4.2   Utilization of red-mud

As mentioned earlier, red-mud is the slimy residue obtained during extraction of alumina from bauxite ore by alkali treatment. It is a heterogeneous material and because of presence of iron it looks red in colour. Chemical composition of red mud generated by Bharat Aluminum Co. (Korba), India, is shown in Table 4.21 below.

Table 4.21 Chemical composition of BALCO (India) red mud

Components	Range (%)
Ferric oxide	36.5–39.2
Silica	6.25–6.55
Alumina(undigested)	16.5–17.5
Calcium oxide	2.0–2.5
Bound soda as Na_2O	5.25–5.60
Soluble/leachable soda as Na_2O	0.55–0.85
Titanium dioxide	18–19
Loss on ignition	9.5

The author has been working with this mud for quite sometime now and product developed as well as philosophy behind these developments are enumerated in detail below.

An examination of the composition of the red-mud given above indicates that from technical point of view presence of large amount of ferric oxide is a hindrance towards utilization of the red-mud. Its occurrence in the red-mud in substantial amount (36–39%), on the other hand, shows that its effective removal will cause concentration of alumina, silica, calcium-oxide and titania to go up almost double. Removal of ferric-oxide is important as it causes early verification (at 1150°C) when heated. This phenomenon in turn causes fall in impact bending strength, generates poor transparency and reduced spalling resistance in the fired product. Further, iron working as a flux, decreases overall strength and iron-silicate layer comes up (floats) during firing. Formation of fired bodies with above composition is complicated as the process does not discriminate phases present in the raw material. Concentration of ferric-oxide, however, in the range 2–2.5% can be tolerated in ceramic/refractory bodies. In any case, removal of ferric-oxide from red mud is an uphill task as the particle size of red mud is in the range of 2–3 micron which makes it all the way difficult to separate from other gangue material. Further ferric oxide being paramagnetic it can not be separated by physical means like Fe_3O_4 which is ferromagnetic. Ores of Fe_2O_3 is called Hematite while that of Fe_3O_4 is called Magnetite for the same reason. However, ferric oxide occurs in two forms – alpha and gamma. While alpha form is ferromagnetic, gamma form is ferromagnetic. The alpha form occurs in nature as hematite while the gamma form is produced by oxidizing Fe_3O_4 or another ferrous compound which is oxidizing so slowly that Fe_3O_4 is first formed. Yellow and red forms of Fe_2O_3 are stable in acids but varieties obtained above 950°C (as with red mud) are insoluble in cold and hot acids (except hydrofluoric acid). In general ion is as stable as ferric, but covalent state ferric is more stable than ferrous. Ferric chloride ($FeCl_3$) is very soluble in water but ferric fluoride (FeF_3) is almost insoluble. $FeCl_3$ is also soluble in acetone and methyl-alcohol. The colloidal ferric oxide, known as "hydrosol" is being prepared by treating solution of ferric chloride with hydrated ferric oxide, added in small quantities at a time, or by adding ammonium carbonate slowly to ferric chloride solution. Ferric ion can be reduced by hydrogen ($357°C$) and later reaction with hydrochloride-acid produces ferrous-chloride ($FeCl_2$) which melts at $672°C$ and boils at $1030°C$. Ferrous chloride is very much soluble in hydrochloric acid and forms complexes like $(FeCl_4)^-$ and $(FeCl_6)^{-3}$. Ferric ion has strong affinity towards oxygen and thus easily reacts with alcoholic –OH group, phenols and ethers. Hematite changes from amorphous to crystalline state at $400–600°C$. Alpha ferric oxide is stable at low temperature while beta-form is stable at higher temperature. Ferric oxide undergoes an irreversible transformation from gamma to alpha form between $40°C$ and $700°C$. Calcined ferric-oxide also does not dissolve completely in aqua-regia. Ferric-oxide however is soluble in phosphoric acid. Ferric oxide (ignited variety) dissolves in hydrochloric acid with great ease and stannous chloride accelerates dissolution of ferric oxide in hydrochloric acid. The ignited variety of ferric oxide readily dissolves in warm 40% sulphuric acid. However, nitric acid dose not react at all with ignited variety of ferric oxide. Aluminum powder reduces ferric oxide to metallic state. Mixture of ferric oxide with chromic oxide forms a continuous series of solid solution. In absence of reducing gases, ferric oxide is not reduced by solid carbon below $950°C$.

4.2.1 Removal of ferric oxide

In particular there are two effective methods for separation of iron from red-mud which can be scaled up to commercial size. These are molten salt reduction and carbon reduction in presence of chlorine gas.

Present author experimented [207] by fusing red mud with potassium-hydrogen-sulfate for removing iron from red-mud. Dried and powdered red mud was mixed with three times of its weight of potassium-hydrogen-sulfate and then fused in the burner at about $800°C$ for $10–15$ minutes. The salt fuses and first shows a greenish colour followed by dark coloration and drying of the mixture. The solid mass after cooling is mixed with concentrated (35%) hydrochloric acid and heated to about $90°C$ for $20–25$ minutes. The solid mass partially dissolves in the acid and after cooling, the residue is washed well with distill water and then oven dried. Composition of this residue is shown in Table 4.22 below. As can be seen in the Table 4.23, the filtrate contains iron in ferric state and titanium as its chloride. This solution or filtrate can be treated by precipitation or by selective electrolysis to separate the metals.

Table 4.22 Chemical composition of the residue

Component	Percent
Titanium dioxide	30
Silica	15
Alumina	34
Calcium oxide	7
Iron oxide	1
Loss on ignition	10

Table 4.23 Chemical composition of the filtrate

Method of analysis	Atomic absorption spectroscopy
Iron	98%
Titanium (as chloride)	2%

Note: All iron could be precipitated by addition of sodium hydroxide solution and can be converted to ferric-sulphate crystals after adding sulphuric acid to the solution.

4.2.2 Forming electroceramic bodies

Electroceramic bodies for its stringent property requirements need use of purest form of raw materials for preparation of its recipe. Electroceramic insulators are used commercially in high voltage tension lines. Their application ranges from 500 to 50000 volts. While the design of the insulator depends basically on its application (like pin insulator, disc insulator, post insulator, etc) its composition is confined within a fixed range of kaolin, feldspar and silica as shown in triaxial diagram below. Besides voltage difference, insulating materials used for high frequency work should have product of dielectric constant and power factor as small as possible. Electroporcelain having "steatite body" has also been used in high voltage lines and because of their relatively lower temperature of formation (compared to electroceramics) and at the same time improve the glass electrical-insulation properties, the effectiveness of depressing effect, i.e. the effect of substituting alkaline-earth oxides for part of silica is found to be pronounced. Electrical insulation properties of toughened glass insulators are of great importance especially when they are used in high voltage transmission lines located in desert regions or in high temperature, high moisture and/or heavily polluted areas. It is for this reason that porcelain or electro-ceramic compositions are chosen for application in high voltage lines where, once applied it has to last for a long period. These bodies are almost vitreous at pyrometric cone equivalence of 10–12. Presence of mica in its constituent raw material causes devitrification of the body. As can be seen in above triaxial diagram, besides steatite bodies, high alumina bodies have also been prepared for electroceramic applications, especially for its superior mechanical strength. Glass compositions, in this respect, are attractive for the reason that it saves energy during its production. However, going from ceramic to glass composition increases possibility of thermal puncture, which is indicated by the "dielectric loss" factor (tan δ). The amount of energy stored in the dielectric at a given field is proportional to its dielectric constant. The product of power factor and the dielectric constant is a measure of the actual loss in the dielectric and is known as "loss factor". The energy lost is itself transformation into heat due to the resistance of the insulating material. It is well known that glass melting temperature increases with increasing amount of silica and alumina, and decreases with an increased substitution of silica by glass modifiers. Generally speaking, substitution of silica by some glass modifiers, such as alkali and alkaline earth oxides will weaken other properties like electrical resistance. But this is not always the case. From the view point of both the origin of the glass structure and its influence on properties, these two apparently contradictory properties (i.e. viscosity and dielectric loss) could be combined perfectly changing the relationship between the glass components. Another basic material CaO has been found to lower PCE value of porcelain bodies by 6 cones. It also lowers modulus of elasticity, increases impact bending strength and coefficient of thermal expansion (spalling resistance is therefore reduced). Zero or even negative coefficient of expansion in porcelain bodies can be achieved by adding small amount of lithium. Cordierite compositions have also been used for this purpose.

For developing hard porcelain insulator body, all that is necessary is to grow as much mullite crystals ($3\ Al_2O_3 . 2SiO_2$) as possible. Amount of mullite again increase with increase in kaolin clay content in the recipe. Thus, a fired porcelain body shows three distinct phases – a vitreous isotropic mixture consisting of feldspathic glass with different degree of saturation by alumina and silica, mullite crystals dispersed in glass (silica-feldspar) phase, and some pores distributed throughout the body. Feldspar in the recipe although melts at 1140°C, its viscosity is very high at this temperature and comes to the range of interaction with clay particles only around 1250°C. Accordingly these ceramic bodies are fired at about dielectric strength. The later phenomenon occurs during stretched heating period when mullite crystals partly dissolve in feldspar glass. Maximization of mullite content is further ensured by restricting iron composition not

Table 4.24 Typical properties of electro ceramics

Property	Values (range)
Compressive strength	4000–5000 kg/cm^2
Tensile strength	250–300 kg/cm^2
Impact bending strength	1.8–2.4 kg/cm^2
Modulus of elasticity	6000–8000 kg/cm^2
Transverse breaking strength	20–60 kg/mm
Resistivity	2.7 × 10^{13} ohm cm
Dielectric constant	6 (air = 1)
Unglazed strength (25 mm rod)	700 kg/cm^2
Glazed strength (25 mm rod)	900 kg/cm^2
Maximum body expansion	11 down (1350°C)

exceeding 2 % in the mixture. Similarly TiO_2 also up to 2 % promotes development of mullite. Here TiO_2 works as a nucleating agent for mullite growth. Further the amount of pores in the ceramic body also needs to be controlled otherwise strength of the porcelain body decreases. Typical properties of porcelain prepared on above guidelines are shown in Table 4.24.

As can be seen in above table, application of glaze (about 0.3 mm thick) increases its strength further by about 25%. While ceramic structure impacts higher compressive strength (less tensile strength), glass structure decreases tensile strength and increases electrical breakdown point. Also higher the glassy phase in the body, less will be its expansion. High alumina bodies have dielectric strength around 25 KV/mm and volume resistivity 10^{14} ohm/cc.

Based on above guidelines, electro ceramic compositions have been developed mainly with four raw materials – ball clay, china clay, feldspar and quartz. While these compositions maintain adequate supply of alumina and silica for optimum growth of mullite and glassy phase, it also provides necessary plasticity for handing in green stage and retaining shape till biscuit firing. Other minute substances may be added, if necessary, to counteract expansion/contraction in final high temperature firing. Finally, it is important to maintain necessary particle size (-200 BS mesh) in order to achieve melting rate.

Based on above information's and the composition obtained after removal of ferric oxide (as shown in Table 4.25 below), raw material composition was adjusted by the authors by addition of china clay and ball clay, basically to bring their ratio to right proportion and conductive for mullite growth.

Table 4.25 Removal of rate of iron from red mud by various reagent

No.	Reagent	% Iron removed
1	Concentrated HCl (with small amount of MnO as accelerator)	35%
2	Chromic acid (freshly prepared, without any catalyst)	55%
3	Fusion with $KHSO_4$ followed by leaching with concentrated acids (90–100°C)	98%

Development of commercial electro-ceramic body composition from above red-mud residue and control of chemical components in it

Concentration of ferric oxide

In standard practice [281] ferric oxide concentration in the green electro ceramic body is never allowed to exceed 2%. The reason being that high concentration of ferric oxide brings down the softening temperature of the body considerably and hinders growth of mullite crystals. The residue composition obtained by the authors in their experiment mentioned above meets this requirement.

Concentration of Titania

In actual practice concentration of titania in the electroceramic body is maintained around 3%. Titania helps in nucleation and growth of mullite bodies. No where in the literature effect of higher concentration of titania is mentioned. In our residue titania concentration is much higher and its effect needs to be studied by designing various experiments in prescheduled time-temperature firing experiments.

Concentration of silica and alumina

These two components are the most important parameters in making finished electro ceramic body. In commercial electro ceramic bodies concentration of alumina and silica is mainly controlled by clay, feldspar, and silica powder as shown in Figure-4.8 (triaxial diagram of the main constituents of electro ceramic body). Feldspar is essential as at high temperature it increases the softening range and thereby increasing chance of growing larger mullite crystals. Ideally concentration of alumina and silica should vary in the range of 2–2.5% to maximize mullite crystal formation but the ratio is fixed by the requirement of both green strength and to compensate high temperature expansion. There are two clays used for this purpose – ball clay and china clay. While ball clay is more responsible for green strength, china clay

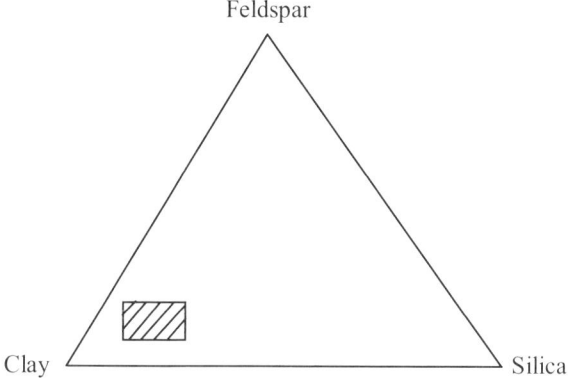

4.8. Triaxial diagram of electroceramic body.

helps forming high temperature bond strength. Thus it is necessary to strike a balance among all these components through repeated experiments and finally fix a composition suitable for electro ceramic body making. In the initial runs as mentioned above, the red mud residue contains favorable ratio of alumina and silica and thus can be wholly consumed in making the body composition with little addition of above clays to make a workable green body composition. Nevertheless, considering other parameters influencing the final body properties, number of high temperature experiments is required for finalizing the optimum recipe.

4.2.3 Making glazes

Glazes while increases compressive strength of the ceramic body (as mentioned earlier), also imparts aesthetic look to the ceramic body. It will be useful here to look at the science behind developing various glazes before we embark upon developing glazes for red-mud based electroceramic body.

First of all we should note that there are both vitrified (glassy) and crystalline variety of glazes available commercially. The crystalline variety receives its crystalline structure from glass composition through repeated heating and cooling cycle. Matte glazes under this variety contain zinc, calcium, and tin or zirconium matte. Other crystalline glazes contain rutile, aventurine, and various colours developed by incorporating liquid or gaseous particles, the refractive index of which is greater or less than the index of the glaze. One such example is the SnO glaze. Glazes are also classified as per their firing temperature. Thus high temperature glazes fired between 1200 and 1400°C are especially used for electro ceramics, bone-china, and bone-porcelain (1250°C) as well as fireclay products (fired between 1200 and 1250°C). Medium temperature glazes are fired at 1000–1050°C and examples are earthenware bodies for tableware, tiles, sanitary wares, etc. The low temperature glazes include majolica (glazes applied to porous bodies). Here we will be particularly interested in high temperature glazes for electro ceramic bodies which are fired in the range 1200–1250°C, which coincides with required firing temperature of the ceramic body itself. Any glaze composition lower than this will cause flow of the glaze during firing, resulting in uneven spread over the surface. Surface tension of the formulated glaze is thus important for its successful use in ceramic bodies. On porcelain bodies it has been found that for glazes high in calcium, strontium, barium or beryllium and highly coloured with chromium, iron or cobalt and similar oxides, the buckling is of convex type, since these glazes hinder shrinkage of the body. Other type of glazes which give the identical fault are those containing a high content of filler material, such as calcined kaolin and those high in alumina. The maximum effect of delaying body shrinkage and hence fostering this type of distortion is obtained with chromium and calcium glazes, and highest body shrinkage effect giving the opposite result is obtained with feldspathic and magnesia glazes.

It is also important to consider while formulating glaze composition that glaze constitution must not be water soluble, for slip is often applied to porous bodies into which soluble part would penetrate and will act as flux to the body while depriving the main glaze of those constituents. However, this does not mean glaze slips do not contain any water soluble material at all. Alkaline nature of some ingredients means that when leaching occurs electrolytes pass into the solution. Clays used in glaze formulation for electroceramic bodies, not only provide necessary silica and alumina in the recipe, but controls viscosity of the prepared glaze for application on ceramic bodies. Viscosity for this purpose can be accurately controlled for these glaze composition based on our knowledge of clay behavior in water as outlined above. Further it may be added here that the action of chemical additives in such glazes is basically on the colloidal particle (normal clays and bentonite) and that finer control can usually be obtained by adding certain amount of such colloidal matters, mainly bentonite type clays. The doping of industrial glazes with bentonite and calcium-chloride is fairly common. Common salt counteracts the swelling of bentonite; a saturated solution will disperse bentonite without causing it to swell. Sodium-chloride concentration up to 2% has this effect. It has been found that bentonite has an enhanced gel forming capacity in the presence of hydroxyl ions which are formed when small amount of MgO are added. Such slips are highly thixotropic. To adjust thixotropic behavior of glaze, ammonium chloride is being added. The usual technique is to add colloidal matter in the form of bentonite, starch, or ball-clay, and then coagulate the slip with ammonium chloride. With some glazes, e.g. those with high clay or kaolin content, it may not be necessary to add the colloid. Ammonium chloride or other flocculants then acts on the colloidal matter already present.

On the other hand, if glaze slip contains too much colloid, it may be necessary to add deflocculant. Sedimentation (e.g. with lead salts) is prevented by increased amount of thioxotrop substance to the system. Galvanized container may cause the glaze to settle by electrochemical reaction. Another reaction which should be considered in preparing glaze slips is the green or confined strength of the dried layer of raw glazes. Where the glaze slip contains up to 10% clay or kaolin, it is often unnecessary to add a small percentage of synthetic organics or binding agent (e.g. cellulose derivatives) which confer green strength to facilitate handling after draining water from the slip. It may be mentioned here that if titania is present, it will develop colour, especially it will influence shades of iron colour.

4.2.4 Utilization of iron filtrate solution

As mentioned earlier, potassium-bisulfate fusion (with BALCO red-mud) followed by extraction with concentration hydrochloric acid of red mud results in a filtrate solution of ferric-chloride which has average composition of 98.5% iron and 1.5% titanium (by AAS) in solution phase [207]. Iron can be precipitated as ferric-hydroxide and titanium can be separated from ferric-chloride solution by solvent (EDTA) extraction. Precipitation of iron by strong sodium-hydroxide solution followed by filtration and drying at 100°C gives rise to a fine powder of ferric oxide, which dissolves readily in dilute sulfuric acid solution to form ferric-sulfate solution. Both these ferric-chloride and ferric-phosphate find application in extracting metals such as lead, uranium, copper, zinc cobalt and nickel by hydrometallurgical route.

Ferric chloride leaching of lead minimizes pollution associated with sulphur dioxide and lead vapor emanation in conventional and cheap smelting extraction route [282]. Ferric chloride oxidation method consists of leaching galena concentrate with ferric chloride-sodium chloride solution to produce lead chloride and lead then extracted by electrolyzing the resultant lead-chloride in a fused salt bath to produce lead metal and the byproduct chlorine gas. The byproduct chlorine is reused to generate ferric-chloride in the leach solution:

$$PbS + FeCl_3 \rightarrow PbCl_2 + 2\ FeCl_2 + S$$

$$2\ FeCl_2 + Cl_2 \rightarrow 2\ FeCl_3$$

$$PbCl_2 \rightarrow Pb + Cl_2$$

Economics vis-à-vis smelting and material corrosion throw challenge for successful adoption of this process to commercial scale. However, the process is simple and straight forward and thus easy to operate and scale-up. The corrosive chloride solution can be handled adequately by a variety of commercially available polymeric materials or titania. Ferric chloride mentioned above controls the overall process economics.

Ferric ion has also been successfully used in [283] heap leaching of uranium at Jaduguda mines of Singhbhum thrust belt (Bihar), India. These ores are basically quartz chloride-sericite with minor amounts of apatite and magnetite having uranium content of 0.06–0.07 U_3O_8. After extraction (dissolution) of uranium in acid medium, uranium values are recovered by ion-exchange and finally precipitated as magnesium diurate [284]. Although conventionally manganese dioxide is used in leaching uranium ores, the acidic ferric-sulphate being a strong and effective oxidizing agent could replace environmentally polluting manganese dioxide as an oxidant in leaching circuit. Here ferric-sulphate in sulphuric acid (50% acid, 4 g sulfuric acid per kg) having pH 1.8 and ferric ion loading rate of 5 g/l with an E_h of −580 mV of the solution is used. With a contact time of 12 hours, the leaching efficiency was found to be 95% from a lean ore of 0.02–0.05% (heap leaching). Particle size of the ore is also important in this respect 84% U_3O_8 values were associated with particle finer than 150 micro-meter (room temperature to 70°C). Uranium extraction efficiency increases from 67% to 90% with decrease in size from 180 micro-meters to 75 micro-meter. Uranium becomes soluble to the acid medium with conversion of U^{+4} to soluble U^{+6}. Associated reaction in this regard may be represented as:

$$UO^{+2} + Fe_2(SO_4)_3 \rightarrow UO_2SO_4 + 2FeSO_4$$

In order to ensure that all iron exists as ferric and not ferrous, other oxidizing agents such as manganese hydroxide, hydrogen peroxide, sodium or potassium chlorates and nitric acid are also added. Ferrous sulphate generated at the completion of leaching reaction, can be converted back to ferric-sulphate either by bacterial (thiobacillus ferrooxidans) action or by chemical means:

$$2FeSO_4 + H_2SO_4 \rightarrow Fe_2(SO_4)_3 + H_2O$$

4.2.5 Extraction of titanium

Next important material needs consideration for economic separation from red mud is titanium from those red muds already rich in titanium content. In particular advantage with these red muds (e.g. BALCO red mud) is that titania content doubles with removal of iron if titania is retained in the precipitate. In the preparation of electro ceramic bodies from this treated red-mud mentioned earlier, titania content to the tune of 2% is tolerable, as at this concentration, it promotes development of mullite crystals which impact great strength to the fired body. Thus titania works as a nucleating agent for millite crystals. At high temperature under strong reducing action titania changes in to its lower form Ti_2O_3 which is dark in colour. Thus in glaze if titania is being used to produce coloures, advantage is taken of this phenomenon to influence shades of iron colour. Red-mud usually softens in the temperature range 800–1200°C and the softening range increases with increase in its alkali contents. Alumina in red-mud can be dissolved by H_2SO_3 or SO_2 to form soluble $Al_2(SO_3)_3$. Titanium in mixed solution of titanium and iron can only be separated when iron is made inactive by sulphidation with H_2S followed by chlorination to form $TiCl_4$. If red mud is heated in 1:4 ratio with sodium hydrogen sulphite (at around 300°C) and the resulting mass digested with sulfuric acid (sp gr 1.89) at 250–300°C, all the aluminum and titanium goes into the solution [285]. Hydrolysis and calcinations of resultant titania results in pigment grade of the oxide. Iron contamination in such processes generally reported around 0.05% and total recovery of titania is around 70%. The sulfuric acid solution of the metal if reduced with iron fillings and the solution after filtration heated to 100°C for 1 hour, H_2TiO_3 precipitates. Prolonged heating with sulfuric acid takes most of iron in solution phase, while most titania remains as precipitate. It may be noted here that procedure followed for extraction of titania from limonite do not suit to extraction of titania from red mud due to occurrence of large amount of alumina in red mud. Mixture of hydrochloride and perchloric acid also reported to yield 80% pure titania with recovery rate of 60–70%. Attempts have also been made to produce ferrotitanium from red mud by using sulfuric acid at 230°C for 1.5–2.0 hours, and precipitating the bivalent metal sulphate (iron sulfate). This water soluble precipitate, when hydrolyzed by boiling with sodium carbonate at around pH = 6 which produces mixed oxide suitable for ferrotitanium metal production. In water medium silica precipitates at pH 2.9 while at pH 4.3 alumina precipitates as aluminum bisulphate from SO_3 treated solution. Titania in red mud (preferably after removing iron by reducing) can be chlorinated to $TiCl_4$ at 650–850°C by prolonged heating for 6–10 hours. If iron is removed, with 98% excess sulfuric acid at even 150°C it can leach out 95% aluminum and titanium. All these dissolved substances then can be precipitated by hydrolysis at 160–180°C. Alkali (sodium hydroxide at 450–500 g/l concentration) when mixed with red-mud and autoclaved at 270–280°C (120 atm pressure), for about 40 minutes 90% of aluminum is leached out. If the alkaline solution of red-mud heated to 180°C for 1 hour, it leaches out both alumina and silica, and residue after filtration when treated with sulfur dioxide gas, it takes away all left-over alumina and silica, leaving behind a residue of titania and ferric oxide. Nitric acid however does not dissolve iron, aluminum and titanium oxides, while rare-earth oxides pass into the solution from which they can be separated by selective ion-exchange separation route.

In recent years carbothermic chlorination process for extracting titanium from its complex has gained momentum for extracting titanium. Chlorination of titanoferrosilicate mass produces titanium-tetrachloride and while all other minerals also chlorinated, silica does not. Titanium-tetrachloride is used in pigment industries, and the process of its manufacture involve carbochlorination of titanium ore, selective condensation and purification of titanium-chloride, oxidation of titanium-chloride vapour to titania pigment in aerosol reactor. The chlorination reaction is being carried out above 900°C, when titanium chloride comes to vapour phase along with other chlorides. Among other chloride contaminants, high melting point ferrous chloride is condensed while high boiling point chlorides like calcium chloride, magnesium chloride, manganese chloride are oxidized with excess oxygen to form a liquid melt with ferrous chloride between 400 and 800°C. The low-boiling chlorides generally oxidize more energetically than iron-chlorides. The kinetic of ferric-chloride formation from ferric-oxide (alpha variety) is slow below 800°C. Sodium chloride catalyzes the oxidation of ferric-chloride gas. However, in fluid bed reactor the conversion is above 95% at 600°C (100 psig pressure) with only 10% excess oxygen. Aqueous hydrochloric acid extract of titanium raw material can also be extracted by pyrohydrolysis in a fluid bed reactor. Pyrohydrolysis in fluidized bed reactor is a well known technology for the treatment of used pickling acids in steel industry. This method in particular is more suitable for removal of iron from ilminite ($FeTiO_3$) which is more abundant than its counterpart rutile (TiO_2). Ilmenite contains 40–60% titania and 14–25% ferric-oxide, while alumina content ranges 0.6–1.0%, silica 0.8–2.2%. However, natural rutile contains 92–95% titania. Possible chemical reactions taking place during carbochlorination of ferric-oxide-titania mixture can be represented as

$$Fe_2O_3 + 3Cl_2 = 2FeCl_3 + 1.5O_2$$

$$Fe_2O_3 + 2Cl_2 = 2FeCl_2 + 1.5O_2$$

$$Fe_2O_3 + 3Cl_2 + 3C = 2FeCl_3 + 3CO$$

$$Fe_2O_3 + 3Cl_2 + 3C = 2FeCl_2 + 3CO$$

$$TiO_2 + 2Cl_2 = TiO_4 + 2O_2$$

$$TiO_2 + 2C + 2Cl_2 = TiCl_4 + 2CO$$

$$TiO_2 + C + 2\ Cl_2 = TiCl_4 + CO_2$$

The carbochlorination reaction initiates when chlorine is slightly deficient than stoichiometric amount. CSIR, South Africa [286] has developed a selective chlorination process whereby the titanium content of the ore is first converted to carbonitride, which subsequently chlorinates at a much lower temperature (typically 350°C) than that required by anove process. Titanium chloride thus produced is reacted with magnesium metal in Kroll process to produce titanium metal. The drawback of Kroll process is that the reaction is vigorous and thus uncontrollable and gives rise of a number of intermediates, while the high volume frothy slag hinders separation of the liberated metal from the rest. In Hunter process these shortcomings were overcome by using sodium metal instead of magnesium. The byproduct sodium chloride is also water soluble and can be easily taken out. If instead of above metal aluminum metal is used, an aluminum–titanium alloy is produced instead of pure metal; this is due to the formation of solid solution at high temperature (800°C) of the reaction.

4.3   Utilization of coal rejects/carbon dust/coal ashes

4.3.1   Utilization of coal rejects and coal dusts

As mentioned earlier, these wastes generated mainly by captive power plants invariable integrated into aluminum smelter plant premises. Composition of these wastes has been shown in Table 14 earlier. As can be seen from these tables, the ash content of these coal wastes ranges from 40 to 80% and they have a wide range of particle size. Coals with such high ash contents can only be best only be best utilized in a fluidized-bed reactor. High ash contents coals when burnt in a fluidized-bed boiler for power generation, it can supplement the load requirement of the smelter plant. Fluid-bed boiler has the advantage of being able to run with as a low as 1% carbon inventory or less than 5% coal in the bed. The combustor temperature is controlled within limits of 750–950°C, the lower limit being fixed on the ignition temperature of the fuel and the upper temperature on the ash fusion temperature of the coal being used. Figure 4.9 below shows a general schematic diagram of a fluidized-bed boiler.

In this regards different boiler configuration has been developed to required steam parameters and quantum of coal available. For example a fluid tube shell boiler with the fluidized combustion chamber located inside the flue tube is offered up to 15 ton/hour steam capacity and for relatively low pressure, while external membrane wall fluidized bed combusting its gases into a shell type boiler to absorb the convective heat forms next grade fluidized boiler in the range 15–30 ton/hour steam capacity. A bi-drum bottom supported natural circulation fluidized boiler is offered for capacities up to 100 ton/ hour steam with medium pressure range (up to 65 atmosphere) and a box type top supported fluidized boiler with horizontal evaporator coil in the bed on forced circulation and the furnace wall on natural circulation is offered

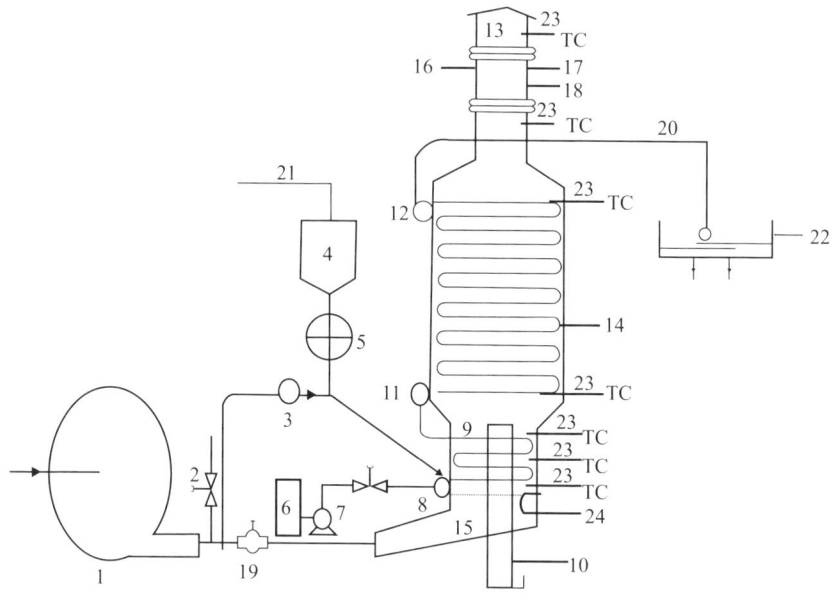

4.9 Schematic diagram of a fluidized bed boiler.

Legend

1. F d fan	13. Flue outlet (chimney)
2. Air by-pass	14. Convection coils
3. Air header for fuel feeding	15. Air box
4. (Coal + lime stone) bunker	16. Dust sampling point
5. Rotary feeder	17. Gas sampling point
6. Soft water tank	18. So$_2$ sampling point
7. Feed water pump	19. Globe value for fluidization air
8. Wall inlet header	20. Hot water outlet
9. Bed coils	21. (coal + lime stone) from mixer
10. Bed material overflow	22. Cooling pond
11. Convection inlet header	23. TC - thermocouples
12. Convection out-let header	24. Distributor plate

for capacities beyond 100 ton/hour steam in both medium and high pressure range. According to authors own experience [287–290] with various grade coals, distributor plate with stand pipe nozzle seems to be appropriate for most of these coal wastes. However, fluidized bed boiler runs on a particular size range and thus calls for sizing the coal wastes to this size range for effective operation of the boiler. Running boiler below this size range will cause more bed elutriation and more carryover in the flue gas of finer particles, when over size will cause inefficient fluidization of the bed and hence lowers its efficiency. Figure 4.10 below shows the typical heat transfer particle size observed in fluidized boilers.

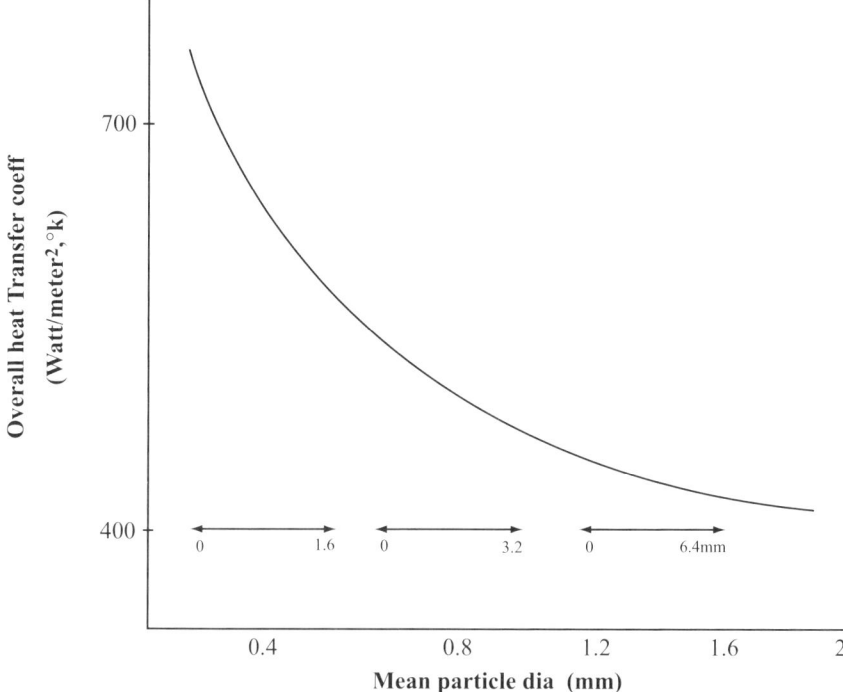

4.10. Variation of heat transfer with particle size in a FBC boiler.

It may be noted in above figure that the coefficient decreases as the particle size increases, while combustion efficiency increases with higher air velocity needed for bigger coal particles. Thus we have to strike a balance these two opposing trend to arrive at an optimal coal size. In general, commercial operation of fluidized bed boiler uses coal with top size of 4 mm with a fluidizing velocity of 2.5 m/s. However, in India power plant crusher output being slightly higher, FBC (fluid bed combustor) boiler manufacturer BHEL (India) chooses 6 mm top size for operation of all their commercial FBC boilers. This is because most of our coals (high ash) being friable, bed attrition results in shift to some extent towards lower size than actual mean feed size during operation. For example, for 2–3 mm top size typical fluidizing velocity chosen is 1 m/s, while that of 4–6 mm size range it is 2 m/s. Coals with still higher friable nature may even allow us to feed up to 5 cm diameter coal particles in a FBC system. Very fine coal or carbon dust may not find use in power generation boilers, but they can be briquette with various binders like pitch fraction, clay, etc, for various industrial/ domestic uses. Here it may be mentioned that although in above discussion we have mentioned only the top sizes of the coal, fluidized boiler actually uses a range of particles when the finer sizes acts as a lubricant to the larger particles and helps in stabilization of the fluidizing condition.

4.3.2   Utilization of coal ash

All aluminum smelter plants have captive power plant within smelter plant premises as current requirement is high for the electrolytic cells. These power houses discharges ash in two forms – fly ash and bottom ash. Fly ash is finer in sizes and is being caught either by a bag house filter system or by a cyclone. Due to very high discharge of fly ash by all boiler houses in the country, its disposal has assumed national importance and task forces has been formed by various governments to tackle and oversee the implementation of fly ash related projects. Major process commercialized for utilization of coal ash includes partial replacement of fine aggregates in asphalt concrete mixtures, geopolymer concrete, brick making, landfill and soil conditioner, in large scale. Heavy metals/radioactivity present in ash is mitigated through cement and concrete industrially which also indirectly reduces emission of green house gas (CO_2) by cement plants [291]. Use of fly ash and bottom ash as partial replacement of fine aggregate in asphalt concrete mixture shows [292] moderate detrimental effect on short time tensile strength. However, it provides adequate minimum tensile strength prescribed by statutory bodies set by US Govt. (Dept. of transportation). Majority boiler ash currently in the USA is being used as landfill. This process is not very economical due to high transport charges. The problem has been mitigated through evolution of technology [293], for example by Separation Technologies LLC, and PMI Ash Technologies LLC, Virginia (USA) unburned carbon was separated from the ash and the residue was used in concrete/cement manufacture. As has been mentioned earlier, coal rejects having low carbon count can also be used in fluidized bed boiler. It may be noted here that using coal ash for making concrete helps in encapsulation of the heavy metals present in ash and thereby forms a very stable matrix which prevents leaching of these toxic metals to the environment.

Fly ash has also been used widely as soil conditioner. These applications include pelletizing fly ash with micro-nutrients including KOH, for slow release of the same to soil, essential for plant growth [294]. Fly ash has been bonded with various binders [295–297] like concrete, clay, etc, to produce durable brick. Readers interested for more information on above subjects are directed to following references:

(a)   Building and engineering material [298–342]
(b)   Soil conditioner [343–354]
(c)   Extraction of heavy metals from coal ashes [355–363]
(d)   Use of coal ash as landfills [364–370]
(e)   Miscellaneous use of coal ash [371–414]

As can be seen in above listings, coal ashes till this date best utilized in making building materials. High rate consumption of these voluminous wastes by these industries equilibrates removal of the wastes from the plant at a stable rate. These building materials include bricks, cements, aggregates, etc. While in the manufacture of cement some compromise has to accept as to the quality of the cement vis-à-vis commercial cements, bricks made from coal ashes seem to command good quality in commercial market. However, coal ash bricks are much heavier than conventional clay bricks and attempts to make hollow bricks to compensate weight has not succeeded much for poorer strength. Because of this high weight, coals ash bricks are not economical beyond about 20 km range from the point of manufacturing base. Fly ash bricks are bonded either by plastic clays or self bonded by application of steam and some cement. Because of above disadvantages coal ashes do not find much taker from the plant and mostly used as landfills or under layers in road construction.

5

Techno-economic feasibility of some selected processes

As indicated in previous chapters, while several processes have been evolved to treat wastes from aluminum plants either for making some industrial products or direct use of these wastes in gross environment management like landfill or making reefs in coast lines or for plantation, still utilization of some waste hazards like spent pot liner or red mud from aluminum plants in large scale remains a challenge . Present authors have worked extensively on such schemes involving economic utilization and safe disposal of spent pot liner from aluminum plants, and economic feasibility of some selective process in this regard will be elaborated in this section below.

First of all in the process of economic evaluation for safe disposal and utilization of a K088 category waste like spent pot liner, we need to take into consideration both the tangible and intangible benefits accrued for such treatment in order to make spent pot liner a non-hazardous waste. For example, as elaborated earlier, till this date there is no commercially viable method for converting all spent pot liner produced by a smelter plant to a non-hazardous waste. Consequently, its storage is a highly expensive proposition and is also not safe in the long run. Thus it has become obligatory to treat spent pot liner by aluminum smelters to dispose it as a safe material. Any successful method adopted towards this end, constitute both non-tangible and tangible benefit related to spent pot liner treatment. According to present author chemical treatment of spent pot liner to make the waste non-hazardous and simultaneously recovering its carbon value contribute to both towards these tangible and non-tangible benefit. In the second stage, conversion of this recovered carbon powder to any useful industrial product pays back fully the initial cost of SPL treatment. Of course it is true that aluminum-production plants which aim solely towards metal production may not charter such waste-treatment process but it may facilitate ancillary downstream industries near the plant site to treat and produce industrial carbon products on a profitable basis. Accordingly, quality and cost of the byproducts obtained by such process of SPL treatment constitute an important economic factor for setting up down stream industries for industrial carbon production and makes the integrated process an economically viable proposition.

Quality of carbon powder obtainable from such chemical treatment of SPL has already been dealt in detail in previous chapters; accordingly we will here see cost of treatment of SPL in terms of kg of byproduct carbon produced. The author has studied the same in two stages first in pilot plant level and then in a semi-commercial scale and the costs calculated on 2008 cost basis under Indian scenario. However the cost calculation may well be compared with present ALCOA practice of disposing SPL as land fill. The later process of landfill costs US$ 440 per ton of SPL to make it inert. Another US$ 60 per ton SPL is required for storing the same. Accordingly, total cost of disposing SPL by above practice comes to about $ 500 per ton SPL or Rs. 25000 per ton SPL. All the following calculations are based on 2007–2008 price line and under Indian conditions.

1 ton SPL per day processing plant

Production capacity = 100 kg SPL/batch
Cost of the plant = Rs. 9.0 lakh
Life of the plant = 5 years
Depreciation cost per day = Rs. 500/day

Running cost

(a) Manpower: three semiskilled worker per shift (total shifts = 3/day)
    Pay of the workers Rs. 420/day
(b) Amount of chemical needed for processing 1 ton SPL = 750 litre acid + 750 litre water (70% acid strength basis).
    Cost of chemicals for treating 1 ton SPL = Rs. 30000/
    Cost of other chemicals for treating 1 ton SPL = Rs. 5000/
(c) Water and electricity charge for treating 1 ton SPL = 580/
    Total running cost = Rs. 36500/ton SPL/day.
    Carbon powder recovered per ton SPL = 700 kg
    Cost of production of byproduct carbon = Rs. 55/kg
    Quality of carbon powder:
    Particle size 20 micron
    Ash content = 5–8% (after column floatation ash content = 2%)

Economics of a 15 ton/day SPL treatment plant

Fixed cost

(a) Capacity of the plant = 15 ton SPL/day
(b) Life of the plant = 20 years
(c) Total cost of the plant = Rs. 115 lakh (please see below for the break-up).
    Glass reactor capacity = 500 liter each (charging rate 300 kg SPL/ batch)

Total cost of the reactors (10 nos.) = Rs. 90.00 lakh
Material handling system = Rs. 16.00 lakh
Filtration unit = Rs. 5.5 lakh
Holding tank (treated water) = Rs. 1.5 lakh
Bubbler = Rs. 1.00 lakh
Precipitation tank = Rs. 1.00 lakh

(d)   Depreciation cost of the plant = Rs 1575/day

Running cost

(e)   Pay for 3 semi-skilled worker = Rs. 300/day
Pay for one Supervisor = Rs. 500/day.

(f)   Cost of chemicals [@ Rs. 14000/100 kg SPL Pilot plant data] = Rs. 14,70,000

(g)   Recovered carbon per day from the plant (@60% yield) = 9000 kg.

Total cost of running the plant = Rs. 1,472,375 (d + e + f)

(h)   Sale value of cromium-hydroxide [@ Rs 4400/100 kg SPL, Pilot plant data].

Final cost of production of decontamination & recovery of byproduct carbon = Rs. 90/kg

References

1. Grjotheim K and Welch BJ, Aluminum smelter technology, Aluminum Verlag GmbH, 1980, Dussdelorf.

2. Kirk RE, et al. Aluminum, Wiley Interscience (Wiley), Encyclopedia of Chemical Technology, 4th ed. Vol. 2, pp. 195, 1991, New York.

3. Kirk, R. E., et al. Aluminum, Wiley Interscience (Wiley), Encyclopedia of Chemical Technology, 4th ed., Vol. 2, pp. 196, 1991, New York.

4. Gerogiogis DI and Ydstie B.E. Ibid, A finite element computational fluid dynamics sensitivity analysis of the conception design of a carbothermic aluminum reactor, pp. 407–414.

5. Secrist DR and Clark JM, Anodes for molten salt electrolysis, US Patent No. 4495049, No. 21, 1983.

6. Aune JA, Frommeyer G, Johansen K, and Sadoway DR, Anode for electrolysis of aluminum, USPTO Application No. 20070209944, dated 9/13/2007.

7. Marshall J. Bruno, Aluminum carbothermic technology comparison to Hall-Heroult Process, Edited by Paul N. Crepeau TMS, Light Metals, pp. 395–400, 2003.

8. Agarwal A, Sahu KK, Pandey BD, Solid waste management in non-ferrous industries in India, Resource, Conservation & Recycling, Vol. 42, No. 2, pp. 99–120, Sept. 2004.

9. Blanche L. Ingram, Determination of fluoride in silicate rocks without separation of aluminum using a specific Ion electrode, Analytical Chemistry, Vol. 42, No.14, pp. 1825, December 1970.

10. Mcquarker NR and Gurney M, Anal. Chem., Vol. 49, pp. 53–56, 1977.

11. Fluxo de Analise no. 365/ALBRAS, Determinacao de Fluor Toatal cm Criolita, 1989.

12. Wood RA, Date LS, and Riley KW, A borate fusion method for determination of fluoride in coal, Fuel, Vol. 82, No.13, pp. 1587–1590, Sept. 2003.

13. ASTM D 3761–96. Standard test method for total fluorine in coal by the oxygen Bomb combustion/ion selective electrode method, ASTM International, 2002, West Consho locker PA.

14. Land disposal restriction: Treatment standards for spent pot liners from primary aluminum reduction (KO88), Federel Register, Vol. 65, No. 134, July 12, 2000, EPA (USA).

15. Mazumder B and Devi SR, Study on fluoride contamination - an experience around aluminum smelter plant, Environmental Science & Engineering, Vol. 5, No.1, pp. 13–18, 2007, Angul (Orissa).

16. Sahu A and Vaishnov MM, Study of fluoride in ground water around the BALCO, Korba area (India), J of Env. Sc & Engg, Vol. 48, No.1, pp. 65–68, Jan 2006.

17. Test method for evaluating solid wastes, Physical/Chemical methods (SW-846), Method 9.010 B, Total & amenable cyanide: Distillation, 3rd Ed, US EPA, 1992 Wash D.C.

18. Test method for evaluating solid wastes, Physical/Chemical methods (SW-846), Method 9.010 B, Cyanide Extraction procedure for solids & oils, 3rd Ed, US EPA, 1992, Washington DC.

19. Hume SM, Anode reactivity-influence of raw material properties, R & D Caerbon Ltd, 2nd Ed, 1999, Sierre, Switzerland.

20. Batista JS and Silveira BI, Influence of sodium content on the reactive of carbon anode. Mat Res., Vol. 11, No. 3, Sept. 2008.

21. Product leaflet (technical specifications of prebaked anodes), M/S. Shiziazhung Green Carbon Products, China & M/S. Dalian Economic & Technology Development Co., China.

22. Mazumder B, A process for the removal of sodium from anode butts generated by aluminum smelter plant, Patent filed with CSIR, Patent Office No. 0226 NF 2003, New Delhi, India.

23. Andrejcak M, Soucy G, Patent review of red mud treatment- Product of Bayer process, Acta Matallurgica Slovaca, Vol. 10, Issue 4, pp. 347–368, 2004.

24. Krishna P, Reddy MS and Patnaik SK, Aspergillus Tubingensis Reduces the pH of the bauxite residue (red mud) amended soils, Water, Air & Air Pollution, Vol. 167, No. 1–4, pp. 201–209, October 2005.

25. An Environmentally Friendly Method for Removing Sodium in Red mud, Chemistry Letters, Vol. 35, No. 11, pp.1278, 2006.

26. Patel R, Utilization of steam to recover the caustic and alumina from red mud slurry, 237th ACS National Meeting (Division of Environmental Chemistry), March 22–26, 2009, Salt Lake City, Uthah.

27. Sahu RC, Patel RK, Ray BC, Utilization of Steam from thermal power plant to recover the caustic & Alumina from red mud slurry, 2008.

28. Pascual J, Corpas FA, Lopez-Beceiro J, Benitez-Guerrero M and Artiaga R, Thermal Characterization of a Spanish red mud, Journal of Thermal Analysis and Calorimetry, Vol. 96, No. 2, pp. 407–412, May 2009.

29. Castaldi P, Silvetti M, Santona L, Enzo S and Melis P, XRD, FTIR and Thermal analysis of bauxite ore processing waste (red mud) exchanged with heavy metals, Clays and Clay Minerals, Vol. 56, No. 4 pp. 461–469, August 2008.

30. Banvolgyi G and Siklosi P, "The improved low temperature digestion (ILTD) process: An economic and environmentally sustainable way of processing gibbsitic bauxites", Light Metals, pp. 45–53, 1998.

31. Harato T et al., Proceedings of the Fourth International Alumina Quality Workshop, pp. 311–320, 1996, Drawin, Australia.

32. Banvolgyi G et al., Banyaszati es Kos. Lapok, Kohaszat, vol. 128, pp. 459–466, 1995.

33. Galarrage RA, Carneiro RR, Keane RE, and Nguyen G, "CVG bauxilum red mud neutralization," Light Metals, pp. 133–137, 2002.

34. Patnaik N, Das B, and Rao RB, "Removal of of calcium and iron oxides from bauxites for use in refractory industry", Erzmetall, Vol. 49, No.9, pp. 555–558, 1996.

35. Rao RB, Reddy BL, and Banerjee GN, "The effect of pre-treatment on magnetic separation of ferruginious minerals in bauxite," Mag. Elect. Separ., Vol. 8, pp. 115–123, 1997.

36. Kpvalenko EP, "Improvement of the process of alumina production at Nikolaec alumina plant", Light Metals, pp. 55–58, 1998.

37. Brown SO and Kirkpatrick DB, "Red mud product development," Light Metals, pp. 25–30, 1999.

38. Pilurzu S, Cucca L, Tore G, and Ullu F, "New research proposals for utilization and disposal of bauxite red-mud from Bayer process," Global Symposium on Recycling, Waste Treatment and Clean Technology, (REWAS'99), Vol. 1, pp. 471–480, 1999.

39. Kirkpatrick DB, Redmud product development, Light Metals, pp. 75–80, 1996.

40. Martinent-Catalot V, Lamerant J-M, Tilmant G, Bacou MS, and Ambrosi JP, "Bauxiline: A new product for various applications of Bayer process redmud," Light Metals, pp. 125–131, 2002.

41. Nunn RF, "Advances in red mud dewatering and disposal technologies," Light Metals, pp. 107–113, 1998.

42. Goldstein GL and Reimers RS, Trace element partitioning and free water bioavailability in synthetic red-mud sediment", Light Metals, pp. 19–24, 1999.

43. Gambrell R and LeBlanc CL, Suitability of red mud amended compost and to toxic degree material for Wetland creation & resstoration. A field study of soil chemical properties and trace and toxic metals release", Draft report from Louisana state University to Kasier Aluminium & Chemical Corporation, 2 Sept. 1998.

44. Ribeiro DV, Labrincha JA and Morelli MR, Chloride Diffusity in red mud added-concretes determined by Migration tests, 11th International Conference on Advanced Materials, ICAM, 2009.

45. Davoodi MG, Nikraz H, Chemical and Physical Characterization of bauxite residue (red mud) for concrete making, Paper presented in Cu 2007, Plenary Session 5 and 6, 2007.

46. Shashikanta M and Agarwal K, Sintering behavior of red mud compact, Submitted B. Tech. Thesis, NIT, Rourkela (India), 6th May 2009.

47. Jobbagy V, Kovacs J, Szeiler G and Kovacs T, Dependence of radon emanation of red mud bauxite processing wastes on heat treatment, Journal of Hazardous Materials, Vol. 172, Issue 2–3, pp. 1258–1263, December 2009.

48. Koleva DA, Copuroglu O, Breugel KV, and de Wit JHW, Electrochemical impedance and potentio-dynamic polarization of construction steel in cement extract, containing red mud and chloride contamination, Report submitted to Delft University of Technology, the Netherlands.

49. Collazo A, Cristobal MJ, Novoa XR, Pena G and Perez A, Electrochemical Imped-ance Spectroscopy as a tool for studying steel corrosion inhabitation in simulated

concrete environment of red mud used as rebar corrosion inhibitor, MC Journal of ASTM International, Vol. 3, Issue 2, February 2006.

50. Zaharaki D and Komnitsas K, Effect of additives on the compressive strength of slag-based inorganic polymers, Global NEST Journal, Vol. 11, No. 2, pp. 137–146, 2009.

51. Utilization of metallurgical solid by-products for the development of inorganic polymeric construction materials, Global NEST Journal, Vol. 11, No. 2, pp. 127–136, 2009.

52. Dimitrios D. Dimas, Ioanna P, Giannopoulou and Dimitrios Panias, Utilization of red mud for synthesis of inorganic polymeric materials, Mineral Processing and Extractive Metallurgy Review, Vol. 30, Issue 3, pp. 211–239, July 2009.

53. Harris M. A., Structural improvement of age-hardened gypsum-treated bauxite red mud waste using readily decomposable phyto-organics, Environmental Geology, Vol. 56, No. 8, pp. 1517–1522, February 2009.

54. Ghosh, P.K., Utilization of red mud and pond ash for construction of embankments, Submitted as BTech Thesis-2009.

55. Ms. Grissett, Why not use red mud and gypsum waste for NOLA Levees? Technical issues by Admin, September 14, 2009.

56. Green M.D., Guingand N.J., deBoger, D.V., Hydrometal, 94 Exploitation of shear and compression rhelogy in disposal of bauxite residue, Pap. Int. Symp., Publisher: Chapman & Hall, Vol. 94, pp. 971–982, 1994, London, UK.

57. Nayak SS, Compression strength of saline water exposed epoxy system containing red mud particles, A Thesis submitted for the degree of B.Tech. in Mechanical Engineering, 2009.

58. Samanta B.C., Maity T., Dalai S. and Banthia A.K, Toughening of Epoxy Resin with solid amine terminated poly (ethylene glycol) benzoate and effect of red mud waste particles, Journal of Materials Science and Technology, Vol. 24, No.02, pp. 272–278, 2008.

59. Samanta B, Effect of modifier and filler concentration on properties of epoxy-red mud waste composite, Paper pre-print, Emerald Group Publishing limited, 2008.

60. Singh B and Gupta M, Surface treatment of red mud and its influence on the properties of particular-filler polyester composites, Bull. Mater. Sci., Vol. 18, No. 5, pp. 603–621, September 1995.

61. Kalkan E, Utilization of red mud as a stabilization material for the preparation of clay liners, Engineering Geology, Vol. 87, No. 3–4, pp. 220–229, 2006.

62. Kolias et al., Stabilization of clay soils with high calcium fly ash hand cement, cement and concrete composites, Vol. 27, pp. 301–313, 2005.

63. Pontikes Y, Angelopoulos GN, Kim U Lee, On the plasticity of clay mixtures with bauxite residue, Poster Presented at the 11th International Ceramic Congress and 4th Forum on New Materials, CIMTEC, Acireale, Sicily, Itley, June 2006.

64. Experts Assessment Meeting of Cement Manufacturing held 17th National Construction Machinery Production Standard Implement in July 2005.

65. Vladimir Cablik, Characterization and application of red mud from bauxite processing, Ph.D. thesis submitted to VSB- Technical University of Ostrava, 2007.

66. Tsakiridis PE, Agatzini-Leonardou S and Oustadakis P, Red mud addition in the raw meal for the production of Portland cement clinker, Journal of Hazardous Materials B, Vol. 116, pp. 103–110, 2004.

67. Tsakiridis PE, Agatzini-Leonardou S, Oustadakis P, Red mud addition in the raw meal for the production of Portland cement clinker, Journal of Hazardous Materials, Vol. 116, No. 1–2, pp. 103–110, 2004.

68. Zhang N, Sun HH, Early-age characteristics of red mud coal gangue cementitious Material, Vol. 1, pp. 6–12, 2009.

69. Van D, Jan SJ, Feng D, Duxson P, Dry mix cement composition methods and systems involving same, U S Patent No. 20090071374, 2009.

70. Zijlstra JJP, Bello V, Collu L, Faux D and Ruggeri R, Passive treatment of AMD with a filter of cemented porous pellets of transformed red mud, Water in minimg Environments, R. Cidu & F. Frau (Eds), IMWA Symposium, 27–31st May 2007.

71. Pontikes Y, Angelopoulos GN, Effect of firing atmosphere and shocking time on heavy clay ceramics with addition of Bayer's process, bauxite residue, Advances in Applied Ceramics, Vol. 108, Issue 1, pp. 50–56, January 2009.

72. Vangelatos I, Angelopoalous GN and Boufoenos D, Utilization of ferroalumina as raw material in the production of ordinary Portland Cement, Hazardoud Materials, 2009.

73. Jitsangiam P, Nikraz H, Jamieson E, Sustainable use of a bauxite residue (red sand) in terms of roadway materials, PhD Thesis Submitted to University of Technology, Perth, Australia, 2007.

74. Desai RD and Peer Mohammed, J. Indian Chem. Soc. (Ind. News Edn.) Vol. 8, pp. 9–13, 1945, Chem. Abst. Vol. 40, pp.3575, 1946.

75. Agatzini-Leonardou S, Oustadakis P, Tsakiridis PE, and Markopoulos CH, Titanium leaching from red mud by diluted sulfuric acid at atmospheric pressure, Journal of Hazardous Materials, Vol. 157, No. 2–3, pp. 579–586, 2008.

76. Swarup D and Sharma AS, Trans Ind. Ceram. Soc. Vol. 4, pp. 75–85 1945; Chem Abstr. Vol. 40, pp. 4857, 1946.

77. Xiang Q, Liang X, Schiesinger ME, and Watson JL, Low temperature reduction of ferric iron in red-mud, Light Metals, pp. 157–162, 2001.

78. Misra B, Staley A, and Kirtepatrick D, "Recovery and utilization of iron from red mud," Light Metals, pp. 161–165, 2001.

79. Liu P, Huo Z, Gu S, Ding J, Zhu J, Liu G, Magnetic dressing iron mineral concentrate from Bayer and red mud, " Light Metals, pp. 149–153,1995.

80. Sahoo KC and Pant A, Preparation of iron ore for use in oxide contents of making, SGAT Bull., Vol.3, No. 1, pp. 43–51, 2002.

81. Mazharenko, N. M. and Noskov, V. A., " Possible directions in the use of red mud in metallurgical production," Metallurgicheskaya I Gornourdnaya Promyshlennost' Vol.2, pp. 127–128, 2001.

82. Tathavadkar V, Antony MP and Jha A, " Improved extraction of aluminium oxide from Bauxite and red mud, " Light Metals, pp. 199–203, 2002.

83. Kasliwal P and Sai PST, "Environment of titanium dioxide in red mud: A kinetic study," Hydrometallurgy, Vol. 53, No.1, pp. 73–87, 1999.

84. Sayan E, and Bayramoglu M, "Statastical modeling of sulfuric acid leaching of TiO$_2$ from red-mud," Hydrometall, Vol. 57, pp. 181–186, 2000.

85. Nikolaev IV, Zakharova VI, Khajrullina RT, "Acid methods of red mud processing: Problems and prospects," Izvestiya Vysshikh Uchebnykh Zavedenii, Tsvetnaya Metallurgia, Vol. 2, pp. 19–26, 2000.

86. Semra Coruh, Immobilization of copper floatation waste using red mud and clinoptilolite, Waste Management & Reasearch, Vol. 26, pp. 409–418, 2008.

87. Dai, Q Spitzer, Paul D, Howard HI and Chen H, Use of silicon-containing polymer to improve red-mud flocculation in the Bayer Process, U S Patent Application 20080257827, 23 October 2008.

88. Visnja O, Karlo N, Vladivoj V, Nenad M, and Olivio M, Red mud and waste base: raw materials for coagulant production, Journal of Trace and Microprobe Techniques, Vol 19, Issue 3, pp. 419–428, 2004.

89. Ghorbani Y, Oliazadeh M and Shahverdi, RIA, Microbiological leaching of Al from the waste of Bayer process by some selective fungi, J. Chem. Eng., Vol. 28, No. 1, 2009.

90. Deng Ri-lie, LI Ke-di, NIE Cheng-rong, Wen Yu huil , LI Hang-cheng, Effect of ameliorants on protective enzyme system of swamp cabbage in cadmium contaminated soil, Journal of Foshan University (Natural science Edition), January 1, 2008, China.

91. Georg W, Ger Pat 803360, 2 April 1951.

92. MikoS, Takahashi H, Kurata AQ and Kawaminami A. Mizu Shori Gijutsu Vol. 14, No. 8, pp. 817–822, 1973, Chem. Abstr. Vol. 80, pp. 73999, 1974.

93. Ho GE, Mathew K and Gibbs RA, Water Res. Vol. 26, No.3, pp. 295–300, 1992.

94. Namasiyam C and Ranganatham K. Res Ind., Vol.37, No.3, pp. 165–167, 1992.

95. Salvador O, Comments on Catalytic applications of red mud an aluminium industry waste: A review, Applied Catalysis B: Environmental, Vol. 84, Issue 3–4, pp. 732–733, 10 June 2008.

96. Shaobin W, Ang HM and Tade M O, Novel applications of red mud as coagulant, adsorbent and catalysis for environmentally benign processes, Chemosphere, Vol. 72, Issue 11, pp. 1621–1635, August 2008.

97 Sushil ZS, Bratra VS, Catalytic applications of red mud, an aluminum industry waste: A review, Applied catalysis B, Environmental, Vol. 81, No. 1–2, pp. 64–77, 2008.

98. Cakici AI, Yanik J, Ucar S, Karayildirim T and Anil H, Utilization of red mud as catalyst in conversion of waste oil and waste plastics to fuel, Journal of Material Cycle and Waste Management, Vol. 6, No. 1 , pp 20–26, 12th March 2004.

99. Ordonez S, Sastre H and Dez FV, Characterization and deactivation studies of sulfided red mud used as catalyst for the hydrodechlorination of tetrachloroethylene, Applied catalysis B, Environmental, Vol. 29, No. 4, Issue 11 pp. 263–273, 12 February 2001.

100. Klopries B, Hodek W and Bandermann F, Fuel, Vol. 69, No. 4, pp. 448–455, 1990.

101. Watkins TE, Drilling, Vol. 14, No. 5, pp. 76–78, 1953.

102. Sho K. Japan Kokoi, 74 09417, 28 Jan 1974.

103. Lambrini VT, Maria Th. Ochsenkuhn-Petropoulou and Leonidas N. Mendrinos, Investigation of the separation of Scandium and rare earth elements from red mud

by use of reversed phase HPLC, Analytical and Bioanalytical Chemistry, Vol. 379, No. 5–6, pp. 796–802, July 2004.

104. Parkh BK and Goldberger WM, US Environmental Protection Agency 600/2–76–301, pp. 143, 1976.

105. Balakrishnan M, Batra VS, J/Hargreaves JS, Monaghan A, Puford ID, Rico JL and Sushil S, Hydrogen production from methane in the presence of red mud –making mud magnetic, Green Chemistry, Vol. 11, pp. 42, 2009.

106. Hargreaves J, "Red mud and methane generate valuable end products", Report: supported by the University of Glasglow Innovation Network which is funded by The Scottish Government, Europe.

107. Paredes JR, Ordonez S, Vega A, Diez FV, Catalytic combustion of methane over red mud based catalysts, Applied catalysis B, Environmental, Vol. 47, No. 1, pp. 37–45, 8 January 2004.

108. Amritphale SS, Anshul A, Chandra N and Ramakrishnan N, A novel process for making radiopaque materials using bauxite- red mud, Journal of the European Ceramic Society, Vol. 27, Issue 4, pp. 1945–1951, 2007.

109. Genc-Fuhrman H., Bregnhoj H., Mc Vonchie D., Arsenate removal from water using sand-red mud columns, Water research, Vol. 39, Issue 13, pp. 2944–2954. August 2005.

110. Apak R., Tutem E., Huj M., Hizal J., Heavy metal cation retention by unconventional sorbents (red muds and fly ashes), Water Res., Vol. 32, pp. 430–440, 1998.

111. Rubinos D.A., Arias M., Diaz- Fierros F. and Barral M.T., Speciation of adsorbed arsenic (V) on red mud using a sequential extraction procedure, Mineralogical Magazine, Vol. 69, No. 5, pp. 591–600, October 2005.

112. Genc- Fuhrman H., Tjell J. C. and McConchie D., Adorption of Arsenic from water using activated neutralized red mud, Environ. & Sci. Technol. Vol. 38, Issue 8, pp. 2428–2434, 9th March 2004.

113. Palmer, S. J., Nothling, M., Bakon, K. H., Frost and Ray L, Thermally activated seawater neutralized red mud used for the removal of arsenate, vanadate and molybdate from aqueous solutions, Colloidal and Surface Chemistry 2010.

114. Bao W., Zhang Z., Ren X., Li F. and Chang L., Desulfurization Behavior of iron-based Sorbent with MgO and TiO_2 additive in Hot coal gas, Energy Fuels, Vol. 23, No.7, pp. 3600–3604, 9th June 2009.

115. Fois E., Lallai A. and Mura G., Sulfur Dioxide absorption in a bubbling reactor with suspension of Bayer red mud, Ind. Eng. Chem. Res., Vol. 46, No. 21 , pp. 6770–6776, 1 August 2007.

116. Process for water treatment, U S Patent no. 7077963, July 18, 2006.

117. Lin C., Maddocks, G., Lin J., Lancaster G., Chu C., Acid neutralizing capacity of two different bauxite residues (red mud) and their potential applications for treaing acid sulfate water and soils, Australian Journal of Soil Reasearch, April 2006.

118. Hanahan C., McConchie D., John Pohl, Robert Creelman Malcolm Clark and Curt Stocksiek, Chemistry of seawater Neutralization of Bauxite Refinery Residues (red mud), Environmental Engineering Science, Vol.21, No. 2, pp. 125–138, March 2004.

119. Coyle, C.M.; Cashion, J.D, Mossbauer effect study of seawater- neutralized red-mud and its adsorption properties, International Symposium on the Industrial Applications of the Mossbauer effect, AIP Conference Proceeding, Vol. 765, pp. 149–153, 2005.

120. Tuazon D. and Corder G.D., Life cycle assessment of seawater neutralized red mud for treatment of acid mine drainage, Resources, Conservation and Recycling Vol.52, pp. 1307–1314, 2008.

121. Menzies W., Fulton I. M., Rosemary A. and P. M. Kopittke, Fresh water leaching of alkaline bauxite residue after sea water neutralization, Neal J. Environ Qual., Vol. 38, pp.2050–2057, 2009.

122. Akcil A., and Koldas S., Acid mine drainage (AMD): causes, treatment and case studies, Journal of Cleaner Production Vol. 14 (12/13), pp. 1139–45, 2006.

123. Varnavas S. P., Boufounos D. and Fafoutis D., An investigation of the potential application of bauxite residue for the improvement of environmental conditions in a marsh environment, Proceeding of the 10 th international conference on Environmental Science and Technology, 5–7 September 2007, Kos island, Greece.

124. Anderson J. D., Bell R. and Phillips I., Amending bauxite residue sands with residue fines to enhance growth potential, National meeting of the American Society of Mining and Reclamation, Gillette, WY, 30 years of SMCRA and Beyond June 2–7, 2007.

125. Udeigwe, K.; Wang, J.; Zhang, and Hailin, Effectiveness of bauxite residue in Immobilizing Contaminants in Manure Amended soils, Soil Science, Vol. 174, Issue 12, pp. 676–686, Dec.2009.

126. Varnavas S. P. and Boufounos D., The potential application of bauxite residue in solid waste stabilization and plant development, Proceeding of the 10 th international conference on Environmental Science and Technology, Chania, 3–5 September 2009, Crete, Greece.

127. Summers R. N., Bolland M.D.A, and Clarke M.F., Effect of application of bauxite residue (red mud) to very sandy soils on subterranean clove yield and response, Australian journal of soil research, Vol. 39, No. 5, pp. 979–990, 2001.

128. Summer R.N., Smirk D.D and Karafilis D.,Phosporous retention and leachates from sandy soil amended with bauxite residue (red mud), Austalian journal of Soil Research, Vol. 34, No. 4, pp. 555–567, 2004.

129. Khan T. A., Ali I., Singh V.V. and Sharma S., Utilization of Fly ash as low cost adsorbent for the removal of methylene blue, malachite green and Rhodamine B dye from textile wastewater, Journal of Environmental Protection Science, Vol. 3, pp. 11–22, 2009.

130. Gupta V. K, Suhas and Ali I., Removal of RhodamineB. Fast green and methylene blue from wastewater using red mud- An aluminum industry waste, Ind. Eng. Chem. & Res. Vol.43, pp.1740–1747, 2004.

131. Allaire, C., Refractory material produced from red-mud, U S Patent No. 5106797, ALCAN INTERNATIONAL, 21 April 1992 Montreal (CANADA) .

132. James G. Hant and Akshay Mathur,Manufacture of ceramic tiles from spent pot lining, U.S.Patent no. 5558690, September 24, 1996.

133. Yalcin, N. and Sevinc, V., "Utilization of bauxite waste in ceramic glass", Ceramics Int., Vol.26, No. 5, pp. 485–493, 2000.

134. Balasubramanian, G., Nimje, M.T., and Kutumbarao, V.V., "Conversion of aluminium industry wastes in to glass- ceramic products," Proceeding Fourth international Symposium on Rexcycling of Metals and Engineered Materials, Warrendale, PA, pp. 1223–1228, 2000.

135. Mahata, T., Sharma, B.P., Nair, S.R., and Prakash, D., Formation of aluminium titanate- mullite composite from bauxite red-mud, Met.Mat. Trans. B, 31B, pp. 551–553, 2000, (USA).

136. Yang J., Zhang D., Hau J., He B. and Xiao B., Preparation of glass- ceramics from red mud in the aluminium industries, Ceramics International, Vol. 34, Issue 1, pp. 125–130, January 2008.

137. Veretnova K.I., Serpeninova T.E & Logvinenko A.T., Izv Sib. Otd Akad Nauk SSSR. Ser.,Khim Nauk, Vol. 120, No. 4, Issue 3; 1973, Chem. Abstr. Vol. 79, 128733q., 1973.

138. Feige R & Dams R. Ger Offen. 2425234,11 Dec. 11 pp., 1975, Chem. Abstr., Vol. 84, 125958j, 1976

139 Aggarwal P.S. & Lele R.V., Indian Ceram. Vol. 18, No. 9, pp.312–15, 1975,; Chem. Abstr., Vol. 86, 7975f, 1977.

140. Chemokomplex Vegyipari Gep-Es Berendezes Export-import Vallalat et al. British Patent no. 1491432, Nov.9, 1977.

141. Akihira S., Yoshikuni T & Tateo H., Japan Kokai 79113611, pp.4, 5 Sept. 1979, Chem. Abstr. 1980, 92,27343a.

142. Licencia Talalmanyokal Ertekesitoe Vallatat, Ger offen. 3002346, 26pp., 13 Aug. 1981, Chem.Abstr. Vol.95, 224537e, 1981.

143. Ratzenbergev H., Fer (East) DD0158028, pp.8, 22 Dec. 1982, Chem.Abstr. Vol. 99, 27028 w, 1983,

144. Sharma R., Kalhatkat S.L & Chowdhury S.K., Indian Patent 151056, 19 Feb. 1983, Chem.Abstr., Vol. 99, 144991r, 1983,

145. Ina Seito Co. Ltd. Japan Kokai J., P59217665, 4 pp., 7 Dec. 1984, Chem. Abstr. Vol. 102, 136647 y, 1985,

146. Knight J.C., Wagh A and Reid W.A. J. Mater Sci. Vol. 21, No. 6, pp. 2179–84, 1986, Chem.Abstr. Vol. 105, 28614p, 1986.

147. Allairw C., Eur. Patent no. Appl. 318305, 11 pp., 31 may 1989, Chem. Abstr. Vol. 111, 101915 k, 1989.

148. Kobayashi M., Jap. Pat. 01201054, 14 Aug. 1989, Chem.Abstr. Vol.112, 124016 k, 1990.

149. Satapathy A., Mishra S. C., Ananthapadmanabhan P.V. and sreekumar K.P., Development of ceramic coatings using red-mud-A solid waste of Alumina plants, The J. of solid waste management, Vol. 33, No. 2, May 2007.

150. Atasoy A., J Therm Anal. Calorim, Vol. 81, pp.357, 2005.

151. Pontikes Y., Vangelatos I., Boufounos D., Fafoutis D. and Angelopoulos G. N., Environmental aspects on the use of Bayer's process bauxite residue in the production of ceramics, 11th International Ceramic Congress and 4th Forum on New Material, 2006, Sicily, Italy.

152. Pontikes Y., Rathossi C., Nikolopoulos P., Angelopoulos G. N.,Effect of firing temperature and atmosphere on sintering of ceramics made from Bayer process bauxite residue, 10th European Ceramic Conference and Exhibition, Berlin, 2007.

153. Potgieter J.H.,. Horne K.A, Potgieter S.S., Wirth W., An evaluation of the incorporation of a titanium dioxide producer's waste material in Portland cement clinker, Mater. Lett. 57(1), 2002, pp. 157–163.

154. Kubota T., Kamiyoshi H. & Yoida S. , Japan Kokai 77133315, 08 Nov. 1977, 2 pp., Chem. Abstr. 1978,89,94006u.

155. Rudolf R., Ger. offen. 2739493, 8 Mar. 1979, 18 pp., Chem.Abstr. 1976, 90, 208944b.

156. Tanaka H., Yamada K., Sasaki M.& Suzuki I., Japan Kokai 79107911, 24 Aug. 1979, pp.5, Chem.Abstr. 1979, 91, 215572c.

157. Tsalafoutas I.A., Yakoumakis E., Manetou A. and Flioni-Vyza A., The diagnostic X-ray protection characteristics of Red mud an aerated concrete based building material, Br. J. Radiol. 71 (1998), pp. 944–949.

158. Smith N. J., Buchanan V. E. and Oliver G.,The potential application of red mud in the production of casting, Material Science and Engineering: A , Vol. 420, Issue 1–2, pp. 250–253, 25 th march 2006.

159. Jones G., Joshi G., Clark M.and McConchi D., Carbon capture and the Aluminuium Industry: Preliminary Studies, Environmental Chemistry, Vol. 3, No. 4, pp. 297–303, 19 July 2006.

160. Zombak D. and Glowry, Mechanism of neutralization of bauxite residue by carbon dioxide, J. Environmental Engg. Volume 135, Issue 6, pp. 433–438, June 2009.

161. David, Phillip K., Kakaria, Vijay K., Method of treating fluoride contaminated wastes, U.S. Patent No. 4735784, 5th April 1988.

162. Gardner O. M, Cheek, R. L., Treatment of aluminum reduction cell linings combined with use in aluminum scrap reclamation, U.S. Patent No. 4927459,22, May 1990.

163. Hittner, Herman J., Nquyen, Q. C., Stabilization of fluoride of spent potlining by chemical dispersion, U.S. Patent No. 5024822, 18th June 1991.

164. Banker, D. B., Brooks, D. G., Cutshall, E. R., Macauley D. D., Strahan, Dennis F., Detoxification of aluminum spent potliner by thermal treatment, lime slurry quench and post-kiln treatment, U.S. Patent No. 5164174, November 17th 1992.

165. Morgenthaler, G. W., Stuthers, Jeffrey L., Carter, G. W, Plasma tourch furnace processing of spent pot liner from aluminum smelters, U.S.Patent No. 5222448, June 29th, 1993.

166. Lindkvist, J. G., Johnswen, T., Method for treatment of potlining residue from primary aluminium smelters, U.S.Patent No. 5286274, 15th February 1994.

167. Lisbina D. F. , Steel K. M., Treatment of spent pot lining for recovery of fluoride values, Electrode Technology Symp., Light Metals 2007, Vol. 4, pp. 843–848.

168. Sims, B. H., Philipp, C. T., Method of recycling industrial waste, U.S. Ptent No. 5496392, March 5, 1996.

169. George W. Mongenthaler, Jeffrey L. Struthers and George W. Carter, Plasma Torch Furnace Processing of Spent Pot Liner from Aluminium Smelters, U.S. Patent No. 5222448, 29 January, 1993.

170. Banker, D. B., Brooks, D. G., Cutshall, Euel R., Macauley, D. D., Strahan, Dennis F., Detoxification of aluminum spent potliner by thermal treatment, lime slurry quench and post-kiln treatment, U.S. Patent No.5164174, November 17, 1992.

171. O'Connor, W. K., Turner, P. C, Addison, G. W., Method for processing aluminium spent potliner in a graphite electrode arc furnace, U.S. Patent No.6498282, December 24, 2002.

172. Philip K., Davis E., Method of treating fluoride contaminated wastes, U.S. Patent No.4735784, 1988.

173. Bell N., Andersen, J.N., Pyrolysis system for processing fluorine containing spent and waste materials,U.S. Patent 4158701, Sept. 12 1978.

174. Bell N., Andersen, J.N., Pyrolysis process for spent aluminum reduction cell linings, U.S. Patent No. 4160808, July 10, 1979.

175. Hall R. N., Boulder and Calif, Process for recovering carbonaceous liquids from solid carbonaceous, U.S. Patent No. 4421603, December 20, (1983).

176. Mansfield K., The spent pot lining treatment and resource recovering project, SPL team (1994–2001), Portland Aluminium Company, (Victoria, Australia).

177. Bell N., Andersen, J.N., Lam H.K., Process for the utilization of waste materials from electrolytic aluminum reduction systems, U.S. Patent No. 4113832

178. Bell N., Andersen, J.N., Modified pyrolysis process for spent aluminum reduction cell linings, U.S. Patent No. 4160809, July 10, 1979.

179. Bernard G. W., Howard W. H., Aluminum electrolytic cell cathode waste recovery, U.S. Patent No. 4355017, 19th October 1982.

180. Goodes, C. G.,Wellwood, G. A., Hayden, Jr. and Howard W., Recovery of fluoride values from waste materials, U.S. Patent No. 4900535, (1990).

181. Process for the heat treatment spent pot lining derived from Hall- Heroult Electrolytic Cells, U.S. Patent No. 5365012 (1994).

182. Silveira BI, Dantas AE, Blasques JE, Santos RK, Effectiveness of cement-based systems for stabilization and solidification of spent pot liner inorganic fraction, Journal of Hazard Mater, Vol.98, No. 1–3, pp.183–902003 March 17,.

183. Cooper B. J., Cooper K.M., Brendan Gerard Cooper and John Joseph Cooper, Treatment of smelting By-Products, U.S. Patent No. 0053973 A1, March 16,(2006).

184. Kasireddy, Bernier V., Kimmerle J.L., Frank M.,Recycling of spent pot linings, U.S.Patent No. 6596252, July 22,2003.

185. Lillo-Rodenas M.A., Amoros C. D. and Solano L. A. , Understanding chemical reactions between carbon and NaOH and KOH An insight in to the chemical activation mechanism, Carbon, Volume 41, Issue 2, pp. 267–275, February 2003.

186. Ignasiak B., Method for spent pot liner processing separating and recycling the products therefrom, U.S.Patent No. 0146440A1 (2004).

187. Bush, F. J., Reclaming spent pot lining, U.S.Patent No. 4889695, 26 December 1989.

188. Bontron J.C.,.Personnet P. B, Lamerant J.M., Process for the treatment of spent pot lining from Hall- Heroult electrolytic cells, U.S.Patent No. 5245116, 14, September 1993.

189. Kaaber H., Mollgaard M., Process for recovering aluminium and fluoride from fluorine containing waste materials, U.S.Patent No. 5558847, 24, September 1996.

190. Cashman J. B., Detoxifying spent aluminum potliners, U.S.Patent No. 6190626, February 20, 2001.

191. J. Besida, Pong K.H., Adrien J., Covey G. H., Donnell T.A.and Wood D. G., Process for treating spent potlining containing inorganic matter, U.S. Patent No. 5939035, August 17, 1999.

192. Barnett R.J., Mezner M. B., Method for recovering fumed silica from spent pot line, U.S.Patent No. 6193944 B1 (2001), 2001.

193. Jenkins, D.H., Recovering of aluminium and fluoride values from spent pot lining, U.S.Patent No. 5352419 (1993), 1993.

194. Method of treating spent pot liner material from aluminium reduction cells, Canadian Patent no. 2314123 (1999).

195. Barnett, R. J., Mezner, M. B., Recovery carbon, silica and alumina from spent pot liner, U.S. Patent No. 6217836, 2001.

196. Robert J. Barnett, and Michael B. Mezner, Recovering chloride and sulphate compounds from spent pot liner, U.S. Patent No. 6231822 (2001).

197. Report from the institute of materials minerals and mining, Aluminium smelting-Reprocessing of spent pot linings, Materials World, Vol. 10, No. 9, pp. 37–37, September 2002.

198. Jenkins, D., Hughes, Recovering of aluminium and fluor spent pot lining, European Patent No. WO/1992/012268, (1992), 1992.

199. Mazumder B., Aluminum wastes Cathodes can now be gainfully used as graphite powder - Minerals & Metal Review (Millennium Special Edition), Vol. 27, No. 8, pp. 35–41, 2000.

200. B. Mazumder, A process for the preparation of mould coating suitable for casting of iron and steel from spent pot liners of aluminum industries, Indian Patent No. 766/DEL/2008.

201. Merg, H., Seng, S., Lu, S. & Valdivieso L. A., Elimination of Cr (VI) from electroplating waste water by electrolysis- Sep Sc & Tech., Vol. 39, No. 7, pp.1501–1518, 2004.

202. Lagrega, M.D., Buckingham, P.L., and Evans, J.C., "Hazardous Waste Management", McGraw-Hill International Edition, 1994, Singapore.

203. Kongsricharoern N. and Polprasert C "Chromium removal by a bipolar electro-chemical precipitation process" , Water Science and Technology Vol. 34, No. 9, pp. 109, 1996, accessed on Aug. 16 , 2001.

204. Niyogi, S., Abraham, T.E. and Ramakrishna, S.V., " Removal of Chromium (VI) ions from industrial effluent by immobilized biomass of Rhizus" , Journal of Scientific & Industrial Reasearch, Vol. 57, pp.809, 1998.

205. Hansen, R.D., Bergman, R.W., Klimpel, R. R., A composition and process for froth floatation of coal from raw coal, European Patent No. WO/1985/005566, December 1985.

206. D. Uzun & M. Gulfen, Dissolution kinetics of iron and aluminum from red mud in sulfuric acid solution, Indian Journal of Chemical Technology, Vol. 14, pp. 263–268, May. 2007.

207. Mazumder B., A process for making red oxide pigment from red mud, CSIR (India) Patent No. 136/NF/2006.

208. Mazumder B., A process for breaking and separating waste emulsion generated by aluminum wire rod mills, Indian Patent No. 735130 dated 51/2004.

209. Mazumder B., A process for the preparation of colloidal-graphite from waste cathode blocks generated by aluminum industries, CSIR (India Patent NO. 999/DEL/2001).

210. Stober W, Fink A, and Bohn E. J Colloid Interface Sc.,Vol. 26, pp.62, 1968.

211. The Chemistry of Silica - R.K. Iler. John Wiley & Sons, pp 336, 1979, N.Y.

212. Jacobson C.A, J. Phys. Chem. Vol.48, pp. 413, 1936.

213. Mazumder B., A process for making fluffy variety of Pyrogenic silica at a relatively lower temperature, Indian Patent No. 006/NF/2002.

214. Mazumder B., A process for the preparation of pencil lead from spent pot liners of aluminum industries, US Patent No. 7217378 B2 (dated 15th May, 2007), European Patent No. EP 1664217B1 dated 15th November, 2006.

215. Mazumder B., A process for making dry cell electrode from spent pot liners of aluminum industries, Indian Patent No. 0395/DEL/2006.

216. Mazumder B., A process for preparation of carbon refractories from spent pot liners of aluminum smelter plant, Indian Patent No. 051/DEL/2003.

217 Brosnan D. A, U.S Patent No. 6471931, October 29th 2002.

218. Hart, L.D 9Ed), Alumina chemical science and technology handbook, American Chemical Soc, ISBN 0-91609433-2 (1990).

219. Bove, F., Composition for plugging blast furnace taphole. USA Patent No. 4022739 (1997).

220. Nagel, Thermoset laminated number, US. Patent No. 2534923, 1950.

221. Halder, D. et. al. Developing a tap hole mass using statistical DOX, American Ceramic Society Bull., Vol.79, 2000.

222. Ruther P, Hochofenstichloch und Rinnenmassen, VDEH seminar Feuerfesttechnologie Teil, Ahrweiler, 1999, Germany.

223. Andreas, Dr.rer.nat., Jurgen O, Dip-Ing. and Laurich (2000) Synthetic alumina raw materials - key elements for innovative refractories, Alcoa Indusrial Chemicals, Europe, Frankfurt, Germany, Vol. 23, No. 62–64, Issue 66, pp. 69–70, 1973.

224. Kenzo T. et. al, Tap hole blocking material for metal melting apparatus, US Patent No. 6281266, 2001.

225. Mazumder B. & Devi S. R, Captive utilization of spent pot liner (as anode making) in aluminum smelter, Journal of Environmental Science and Engineering (in progress for publication).

226. Goyal, M., et al., Influence of carbon oxygen surface group on the adsorption of polar and non-polar vapors by activated carbon, Indo-Carbon 2006, HEG-Mandeep, Bhopal (India), pp 55–62, 9–10 Nov, 2006.

227. Bansal, R., Surface groups in activated carbon, Carbon 18, New Delhi (N.P.L), pp. 297, 1980.

228. Boehm, H.P., Advances in catalysis 16, Academic Press, New York, pp. 179, 1966.

229. Sorg,T.V and Logsdon, G.S., Surface absorption of chemicals on carbon, Publisher – American Water Works Association (U.S.A), pp. 411, 1980.

230. Corapciglu, M.O and Huang, C.P., Adsorption of heavy metals on to hydrocarbon activated carbon, Water Research, Vol. 21, No.9, pp. 1031–1044, 1987.

231. Huang,C.P and Wirth, P.K., Activated carbon for the treatment of Cd(II) waste water, J Env Engg (Divn of Am Soc for Civil Engrs), pp.1206, 1982.

232. Hohl, C.P and Stumn, N., Interaction of Pb with hydrous $Y\text{-}Al_2O_3$, J Colloid Surface Science, Vol. 55, pp. 281, 1976.

233. Schindler, P.N., Surface complexes of oxide-water interface, Arbor Science, Ann Arbor, 1981, Michigan (USA),.

234. Snoeynik, V.L and Webr, W.J., The surface chemistry of active carbon – A discussion of structure and surface functional groups , Env Sc & Tech, Vol. 1, pp.228, 1967.

235. Puri, B.R., Inorganic Chemistry & Physics of carbon (Walker P.L edited), Marcel Dekker Inc, New York, pp 248–249, 1970.

236 Joogivesi, Uldnouded, Esti standard, EVS 663, 1995.

237. Lalumandier, J.A & Jones, J.L., Fluoride concentrations in drinking water, J. AWWA, Vol. 91, pp. 42–52, 1999.

238. Veressinina Y., Trapido M., Ahelik V. and Munter R., Fluoride in drinking water: the problem and its possible solutions, Proc. Estnian Acad. Sci. Chem., Vol. 50, No. 2, pp. 81–88, 2001.

239. Campbell A.D., Determination of fluoride in various matrics, Pur & Appl. Chem., Vol. 59, No. 5, pp. 695–702, 1987.

240. Barnett, R.J., and Mezner, M.B., Method of treating spent pot liner material from aluminum reduction cell, Canadian Patent No. 2314123 (2006).

241. Jenkins D. H., Recovery of aluminum and fluoride values from spent pot lining, U.S. Patent No. 5352419, October 4, 1994.

242. Coope B J., Cooper K.M., Coope B. G., and Cooper J.J., Treatment of smelting By-Products, U.S. Patent No. 0053973A1, March 16, 2006.

243. Morse M. E., EIMS Metadata Report – Project, A chemical process to treat spent potliners and produce several recyclable commercially valuable products- phase 1 (68D40056), 5th march 2007.

244. Ingnasik B., Method for spent potliner processing separating and recycling the products therefrom, U.S. Patent No. 0146440A1, July 29, 2004.

245. Goodes C.G., Wellwood G.A., Hayden H.W., Recovery of fluoride values from waste materials, U.S. Patent No. 4900535, February,13, 1990.

246. Besida J., Pong K.H., Adrien J., Covey H, O'Donnell T.A. and Wood D.G., Process for treating spent pot lining containing inorganic matter, U.S.Patent No. 5939035, August 17, 1999.

247. Cashman J. B. Detoxifying spent pot liners, U.S. Patent No. 6190626, February 20, 2001.

248. Bontron J. C., Barrillon E., Personnet P., A process for heat treating spent pot linings from Hall-Heroult electrolytic cells, U.S. Patent No. 5365012, November 15, 1994.

249. Bontron J. C., Personnet P.B., Lamerant J.M., U.S. Patent No. 5245116, September 14, 1993.

250. Bush, F. J., Reclaiming spent pot lining, U.S. Patent No. 4889695, 26th December 1989.

251. Kasireddy V., Bernier , J., Kimmerle, F.M., Recycling of spent Pot linings. U.S. Patent No. 6596252, July 22 2003.

252. Banker, D,B., et al., Detoxification of aluminum spent pot liner by thermal treatment, lime slurry quench and post-kiln treatment, U.S. Patent No. 5164174, November 17, 1992.

253. Mongenthaler, G.W., et al., Plasma torch furnace processing of spent pot liner from aluminum smelters, U.S. Patent No. 5222448, Jun 29 1993.

254. Cashmann, J.B., Detoxifying spent aluminum pot liners, U.S. Patent No. 6190626BI, February 20, 2001.

255. Henning, K. and Mognes, M, Process for recovering aluminum and fluoride from fluorine containg waste materials. U.S. Patent No. 435279, 5th May 1995.

256. Besida, J., Pong, T.K., Adrien, R.J., Covey, G.H., O'donnell, T.A., Wood, D.G., Process for treating spent pot lining containing inorganic matter, U.S. Patent No. 5939035, 17th August 1999.

257. Barnett, R.J., and Mezner, M.B., Recovering chloride and sulphate compounds from spent pot liner, U.S. Patent No. 6231822, 17th April 2001.

258. Lisbona D.F., and Steel K.M., Treatmrnt of spent pot liner for recovery of fluoride values, Light metals, Vol. 4, pp. 843–848, 2007.

259. Jenkins, D.H., Recovery of aluminum and Fluoride from spent pot lining, European Patent No. 012268, 23rd July 1992.

260. Jenkins D. H., Recovery of aluminum and fluoride values from spent pot lining, U.S.Patent No. 5352419, October 4, 1994.

261. Aiso H., Takemura T., Takeuchi T., Process for manufavturing aluminum fluoride, U.S.Patent No. 3855401, December 17, 1974.

262. Pong T.K., Adrien R.J., Beside J., O'Donnell T.A. and Wood D.G., Spent potlining-A Hazardous Waste Made Safe, Process Safety and Environmental Protection, vol. 78, Issue:B3, pp. 204–208, May 2000.

263. Hisao Kurosaki, Reduction of fluorine- containing industrial waste using aluminum-solubility method, Oki Technical Review, Vol.63, pp. 53–56, January 1998.

264. Orth G. O., Jr.; Orth R. D., Recovery of sodium fluoride and other chemicals from spent carbon liners, U.S.Patent No. 4113,831, September 12, 1978.

265. Mongenthaler, G.W., et al., Plasma torch furnace processing of spent pot liner from aluminum smelters, U.S. Patent No. 5222448, 1993.

266. Barnett, R. J. and Mezner M.B., Method of treating spent potliner material from aluminum reduction cells, U.S. Patent No. 4927459, March 3, 1998.

267. Barnett, R.J., and Mezner, M.B., Method of treating spent potliner material from aluminum reduction cells, U.S. Patent No., 6123908, September 26, 2000.

268. Barnett, R.J., and Mezner, M.B., Method of treating spent pot liner material from aluminum reduction cells, U.S. Patent No. 5955042, September 21, 1999.

269. Barnett, R.J., and Mezner, M.B., Method of trating spent potliner material from aluminum reduction cells, U.S. Patent No. 5723097, March 3, 1998.

270. Bell N., Andersen J. N., .Lam H.K. H., Process for the utilization of waste materials from electrolytic aluminum reduction system, U.S. Patent No. 4113, 832, September 12, 1978.

271. Cutshall, E.R., et al., Re-cycling of spent pot liner. U.S. Patent No. 4784733, 1988.

272. Brosnan, D., Process for recycling spent pot liner, U.S. Patent No.6471931, 2002.

273. Hall R.N., Process for recovering carbonaceous liquid from solid carbonaceous particles, U.S. Patent No. 4421603, 1983.

274. Nakajima S., et al., Semiconductor device and its manufacturing method, U.S. Patent No. 7129522, 1998.

275. Bontron J., Barrillon E., Personnet, P. A process for the heat treatment of spent pot linings derived from Hall- Heroult electrolytic cells which comprise carbon and silico- aluminous, U.S. Patent No. 81198, 1993.

276. Sims B.H., et al., Methods of recycling industrial waste, U.S. Patent No. 5496392, 1996.

277. Bontron J., Personnet P., Lamerant J., Process for the treatment for the wet treatment of spent pot lining from Hall- Heroult electrolytic cell, U.S. Patent No. 5245116, 1993.

278. Crnojevich R., Case, A., Rando, F.D., Sweeney, J.D., Recovery of Chromium in high purity state from waste material of etching operation, U.S. Patent No. 5171547, 1992.

279. Silveria B.I., Dantas, A.E., J.E. Blasques, Santos, R.K., Effectiveness of cement based systems for stabilization and solidification of spent pot liner inorganic fraction, J. Hazardous Mater, 98, 183–190, (2003).

280. Bockman, Process for the removal of impurities in reacted alumina, U.S. Patent No. 5251818, June 25, 1985.

281. Mazumder B., Management of R & D center of a large electroceramic industry, Transactions of the Indian Ceramic Socity 43(2), 50–52, 1984.

282. Wong M.M, Haver F.P. & Sandberg R .G, Ferric chloride leach electrolysis for production of lead- (U.S. Bureau of Mines), lead-Zinc-Tin-80, Proc. World Symp. Las Vegas, pp. 445–554, 1980, Nevada, USA.

283. Mathur AK, Vishwamohan K, Mohanty K B, Murty V.K & Seshadrinath S.T. Technical Note:Uranium extraction using biogenic feric-sulfate a case study on ore from Jaduguda, Bihar, India- Mineral Engg 13(5), 575–579, 2000.

284. Rich experience for further challenges: Natl symp. Uranium Technology, BARC (II), pp. 431–462,1989.

285. Schlutz R.E & Maushgen H. Ger patent 2551380, 26th May 1977, 7 pp.

286. Bhatnagar SS, Parthsarathy S, Single GC & Rao S AL; J. Sc. Ind. Res 4, 378 (1945).

287. Borah R.C., Mazumder B. & Bora M.M. Atmospheric fluidized bed combustion of high sulfur high volatile N.E coals of India - Research and Industry, Vol. 40, pp. 315–321,1995.

288. Borgohain J.N. & Mazumder B., Efficient utilization of North-Eastern coals for power generation: Fluidbed technique - J Assam Sc Society (Special Issue) , March 1988.

289. Mazumder B. & Borah R.C., Design of a fluid bed combustor with secondary air injection nozzle for reduction of carbon monoxide level - Patent filed with CSIR (New Delhi) Patent Office, 1991.

290. Mazumder B., Development 2 ton/hr (steam) fluidized bed boiler using high sulfur north eastern coals of India Final project report submitted to North Eastern Council, Shillong (1991).

291. Naik T., Use of fly ash as particle replacement of cement in concrete, reduces in turn, emission of CO_2 by cement plant, University of Milwankee, Wisconsin, USA.

292. Churchill E.V and Amirkhanian S.N., Coal ash utilization in asphalt concrete mixtures, J. of Materials in Civil Engg. Vol. 11, No. 4, pp. 295–301, Nov. 1999.

293. Report to Congress- Wastes from combustion of fossil fuel, Vol. II, EPA 530–S–99–010, Office of Solid Wastes, U.S.EPA, March 1999.

294. Yoo J.G & Jo J.M., Utilization of coal fly ash as a slow release grammar medium for soil improvement, J. of Air & Waste Management Association, Jan. 2003.

295. Sarkar R., Singh N. and Das S.K., Effect of addition of pond ash and fly ash on\ properties of ash-clay burnt bricks, Waste Management & Research, Vol. 25, No. 6, pp. 566–571, 2007.

296. Rostami H. and Solomon S., High performance alkali ash material, J. of Solid Waste Technology and Management, Vol. 30, No. 3, August 2004.

297. Shuvalov Ju. V., Nifontov Ju. A., Ehkgardt V. I., Benin A.A., Nikulin A.N., Method of production of coal bricks, Russian Patent No. 2227803, May 25, 2004.

298. AinetoM., Anselmo A.D., Rincon J.M., Maximina Romero, Production of lightweight aggregates from coal gasification fly ash and slag, Aggregates 2, World of coal Ash Utilization(WOCA) Conference, Lexington, KY,USA, May 4–7, 2005.

299. Camoes A., Aguiar B., Jalali S., Durability of low cost high performance fly ash concrete, Concrete 2, International Ash Utilization Symposium, Center for Applied Energy Research, University of Kentucky, Paper - 43, 2003.

300. Kwang-suk You, Nam-il UM, Gi- Chun Han, Hee Chan Cho and Ji- Whan Ahn, Concrete and concrete IV, World of Coal Ash Utilization(WOCA) Conference, Lexington, KY,USA, May 4–7, 2009.

301. Diaz E.I., Allouche EN. and Eklund S., Evaluation of fly ash stockpiles as potential source material for Geopolymer concrete, Cement and Concrete II, World of Coal Ash Utilization (WOCA) Conference, Lexington, KY,USA, May 4–7,2009.

302. Amaya P. J, Amaya A. J., The use of bottom ash in the design of dams, Aggregates/ Geotechnology II, World of coal ash Utilization (WOCA) Conference, Lexington, KY,USA, May 4–7,2007.

303. Namagga C., Atadero R. A., Optimization of fly ash in concrete: high lime fly ash as a replacement for cement and filler material, Cement and Concrete IV, World of Coal Ash Utilization (WOCA) Conference, Lexington, KY, USA, May 4–7, 2009.

304. Liu H., Banerji S.K., Burkett W. J. and Engelenhoven J. V., Construction, World of Coal Ash Utilization (WOCA) Conference, Lexington, KY,USA, May 4–7,2009.

305. Mohammed A., Elsageer, Millard S. G. and Barnett S. J., Strength Development of concrete containing coal fly ash under different curing temperature conditions, Poster, World of Coal Ash Utilization (WOCA) Conference, Lexington, KY,USA, May 4–7,2009.

306. Barr G., Grerra E., Fly ash cement kilns: green practices leading to emission reduction, Cement and Concrete III, World of Coal Ash Utilization (WOCA) Conference, Lexington, KY,USA, May 4–7,2009.

307. Batmunkh N., Ishida T, Nikraz H., Performance evaluation of coal ash concrete as building material in Mongolia, Cement and Concrete VII, World of Coal Ash Utilization (WOCA) Conference, Lexington, KY,USA, May 4–7,2009.

308. Beck D., Kraemer W., Richard Mack R., Bottom ash use in utility joint trench operation, Bulk3, International Ash Utilization Symposium, Center for Applied Energy Research, University of Kentucky, Paper - 21, 2003.

309. Bhatty J.I., Gajad J., Miller F.M., Commercial decontamination of high-carbon fly ash technology in cement manufacturing, Concrete 1, International Ash Utilization Symposium, Center for Applied Energy Research, University of Kentucky, Paper - 38, 2003.

310. Bhatty J.I., Gajad J., Miller F.M., Conversion of coal prewaste into Portland cements, Novel Applications1, International Ash Utilization Symposium, Center for Applied Energy Research, University of Kentucky, Paper - 39, 2003.

311. Chou M.M., Chou S.F.J., Patel V., Lewis H.S., Kimlinger J. P., Bryant M.M. and Botha F., Commercialization of fired paving bricks with class F fly ash from Illinois basin coals, Posters, World of Coal Ash Utilization(WOCA) Conference, Lexington, KY, USA, May 4–7,2005.

312. Chou M.M., and Botha F., Manufacturing commercial bricks with Illinois coal fly ash, Bulk 2, International Ash Utilization Symposium, Center for Applied Energy Research, University of Kentucky, Paper - 25, 2003.

313. Majkrzak G.L. II, Watson J.P., Bryant M.M., Kip Clayton, Effect of cenospheres on fly ash brick properties, Emerging Technology II, World of Coal Ash Utilization (WOCA) Conference, Lexington, KY, USA, May 4–7, 2007.

314. Bumrongjaroen W. and Livingston R.A., A Figure of merit for fly ash replacement of Portland cement, Cement and Concrete III, World of Coal Ash Utilization (WOCA) Conference, Lexington, KY, USA, May 4–7,2009.

315. Liu H, Banerji S.K., Burkett W.J.and Vanzngelenhoven J., Environmental properties of fly ash bricks, Construction, World of Coal Ash Utilization (WOCA) Conference, Lexington, KY,USA, May 4–7,2009.

316. Mackos R., Butalia T., Wolfe W. and Walker H.W., Use of lime activated class F fly ash in the full depth reclamation of asphalt pavements: Environmental aspects, Aggregates/ Geotechnology I, World of Coal Ash Utilization (WOCA) Conference, Lexington, KY,USA, May 4–7,2009.

317. Wolfe W., Butalia T.S. and Walker H., Full depth reclamation of asphalt pavements using class F fly ash, Aggregates/ Geotechnology I, World of Coal Ash Utilization (WOCA) Conference, Lexington, KY,USA, May 4–7,2007.

318. Nugteren H.W., Butsellerorthlieb V.C.L., Izquierdo M., Witkamp G.J. and Kreutzer M.T., High strength Geopolymers from fractionated and pulverized fly ash, Cement and Concrete VIII, Aggregates/ Geotechnology I, World of Coal Ash Utilization (WOCA) Conference, Lexington, KY, USA, May 4–7, 2009.

319. Crouch L.K., Hewitt R., Byard B., High volume fly ash concrete, VII, Aggregates/ Geotechnology I, World of Coal Ash Utilization (WOCA) Conference, Lexington, KY, USA, May 4–7, 2007.

320. Szczygielski T., Myszkowska A., Coufal R., Kopczynska O., Peat consolidation with bottom ash-theory and practice, Agriculture 1, World of Coal Ash Utilization (WOCA) Conference, Lexington, KY, USA, May 4–7, 2005.

321. Cross D., Stephens J. and Vollmer J., Structural applications of 100 percent fly ash concrete, Cement and Concrete 5, World of Coal Ash Utilization (WOCA) Conference, Lexington, KY, USA, May 4–7, 2005.

322. Dockter B.A., Eylands K.E. and. Hamre L.L., Use of bottom ash and fly ash in rammed-earth construction, Construction and evolution of CCPs, International Ash Utilization Symposium, Center for Applied Energy Research, University of Kentucky, Paper – 56, 1999.

323. Querol X., Umana J.C., Plana F., Alastuey A., Lopez-Soler A., Medinaceli A., Valero A., Domingo M. J., Rojo E.G., Synthesis of zeolites from fly ash in a pilot plant scale: examples of potential environmental applications,Chemistry and Mineralogy, International Ash Utilization Symposium, Center for Applied Energy Research,University of Kentucky, Paper – 12, 1999.

324. Font O., Querol X., Plana F., Lopez-Soler A., Chimenos J.M., March M.J., Esciell F., Burgos S., Pena F.G., Alliman C., Occurance and Disribution of valuable metals in fly ash from pueertollano IGCC power plant, Spain, Chemistry and Mineralogy I, International Ash Utilization Symposium, Center for Applied Energy Research,University of Kentucky, Paper – 98, 2001.

325. Faith T. and Umit A., Utilization of fly ash in manufacturing of building bricks, Construction products II, International Ash Utilization Symposium, Center for Applied Energy Research, University of Kentucky, Paper – 13, 2001.

326. D. Fernandes I.D., Ferret L., Khahl C.A., Endres J.C.T., Maegawa A., Crystalline microstructure modification of Brazilian coal ash with solution, Chemistry and Mineralogy, International Ash Utilization Symposium, Center for Applied Energy Research,University of Kentucky, Paper – 88, 1999.

327. Kim J., Yoon M., Jung S., Han S., Lim N., The Mechanical properties of coal ash generated in ZSouth Korea for using highway road material, Posters, World of Coal Ash Utilization (WOCA) Conference, Lexington, KY,USA, May 4–7, 2009.

328. Kayali O. and Kim A.G., High performance bricks from fly ash, New Products1, World of Coal Ash Utilization (WOCA) Conference, Lexington, KY,USA, May 4–7, 2005.

329. Koukouzas N., Papayianni I., Tsikardani E., Papanikolaou D., Ketikidis C., Greek fly ash as a cement replacement in the production of paving blocks, Cement and Concrete III, World of Coal Ash Utilization (WOCA) Conference, Lexington, KY, USA, May 4–7, 2007.

330. Pimraksa K., Wilwlm M., Kochberger M. and Wruss W., A New approach to the production of bricks made of 100 % fly ash, Construction products I, International Ash Utilization Symposium, Center for Applied Energy Research, University of Kentucky, Paper – 84, 2001.

331. Itskos G.S., Itskos S. and Koukouzas N., The effect of the practice size differentiation of lignite fly ash on cement industry applications, Posters, World of Coal Ash Utilization (WOCA) Conference, Lexington, KY, USA, May 4–7, 2009.

332. LavA.H., M. Lav A., Goktepe B., Analysis and design of a stabilized fly ash as pavement base material, Cement and Concrete 10, World of Coal Ash Utilization (WOCA) Conference, Lexington, KY, USA, May 4–7, 2005.

333. Li L., Edil T.B. and Benson C.H., Properties of pavement geomaterials stabilized with fly ash, Aggregates/ Geotechnology I, World of Coal Ash Utilization (WOCA) Conference, Lexington, KY, USA, May 4–7, 2009.

334. Sutton M.E., Schmaltz T., Miller E.C., Haeper K.J., Radon emission from high volume coal fly ash structural fill site, Environmental Benefits II, International Ash Utilization Symposium, Center for Applied Energy Research, University of Kentucky, Paper – 91, 2001.

335. Mishulovich A., Evanko J.L., Ceramic tiles from high carbon fly ash, Novel applications 1, International Ash Utilization Symposium, Center for Applied Energy Research, University of Kentucky, Paper – 18, 2003.

336. Sato A., Nishimoto S., Effective reuse of coal ash as civil engineering material, Aggregates 2, World of Coal Ash Utilization (WOCA) Conference, Lexington, KY, USA, May 4–7, 2005.

337. Ohnaka A., Hongo T., Ohta M. and Izumo Y., Research and development of coal ash granulated material for civil engineering applications, New products 1, World of Coal Ash Utilization (WOCA) Conference, Lexington, KY, USA, May 4–7, 2005.

338. Izquierdo M., Querol X.1, Phillipart C., Antencc D., Influence of curing conditions on geopolymer leaching, Environmental IV, World of Coal Ash Utilization (WOCA) Conference, Lexington, KY, USA, May 4–7, 2009.

339. Philips B.L., Groppo J., Peronne R., Evaluation of processed bottom ash for use as lightweight aggregate in the production of concrete masonry units, Aggregates 1, World of Coal Ash Utilization (WOCA) Conference, Lexington, KY, USA, May 4–7, 2005.

340. Tempest B., Sanusi O., Ogunro J.G.V., David Weggel, Commpresive strength and embodies energy optimization of fly ash based geopolymer cement concrete, Cement and concrete VIII, World of Coal Ash Utilization (WOCA) Conference, Lexington, KY, USA, May 4–7, 2009.

341. Swan C.W., The use of synthetic lightweight aggregates as a component of sustaibable design, Aggregates/ Geotechnology V, World of Coal Ash Utilization (WOCA) Conference, Lexington, KY, USA, May 4–7, 2009.

342. You K., UM N., Han G., Cho C. H. and Ahn J.W., Manufacturing of cementatitious materials with coal combustion bottom ash and FGD gypsum, Cement and concrete IV, World of Coal Ash Utilization (WOCA) Conference, Lexington, KY, USA, May 4–7, 2009.

343. Thenoux G., Acevedo , Loreto, Osorio A., Gonzalez M., Improved performance of soils stabilized with FBC ash, Aggregates/ Geotechnology III, World of Coal Ash Utilization (WOCA) Conference, Lexington, KY,USA, May 4–7, 2009.

344. Amonette J.E., Kim J., Russell C.K., Palumbo A.V. and Daniels W.L., Enhancement of soil carbon sequenstration by emendment with fly ash, CUBs and sustainable development, International Ash Utilization Symposium, Center for Applied Energy Research, University of Kentucky, Paper – 47, 2003.

345. Buck J.K., CPSSc and LaBuz L.L., Bottom ash fines as a soil amendment for Turfgrass site course laboratory and and mesocom studies at PPL burner Island and Montour steam electric station, Agriculture 3, World of Coal Ash Utilization (WOCA) Conference, Lexington, KY,USA, May 4–7, 2005.

346. Yunusa I. A. M, Eamus D., De Silva D.L., Murray B. R., Burchett M.D., Skilberk G.C. and Heidrich C., Prospects for coal ash in the management of Austelian Soils, Agriculture 2, World of Coal Ash Utilization (WOCA) Conference, Lexington, KY, USA, May 4–7, 2009.

347. Ghiani M. C.R., Peretti R., Zucca A. S.A., Heavy metal immobilization using fly ash contaminatedby mine activity, Mining I, International Ash Utilization Symposium, Center for Applied Energy Research, University of Kentucky, Paper –6, 2001.

348. Skodras G., Karangelos D., Anagnotakis M., Grammelis P. H. E., Kakaras E., Mineralogy and Geochemistry of Greek and Chinese coal fly ash : research for potential applications, Chemistry 5, World of Coal Ash Utilization (WOCA) Conference, Lexington, KY,USA, May 4–7, 2005.

349. Kruger R.A., Surnidge A.K.J., Predicting the efficacy of fly ash as a soil ameliorant, Agrriculture I, World of Coal Ash Utilization (WOCA) Conference, Lexington, KY, USA, May 4–7, 2009.

350. Surridge A.K.J., Merwe V., Kruger R., Preliminary microbial studies on the impact of plants and South African fly ash on amelioration of crude oil polluted acid soils, Agriculture I, World of Coal Ash Utilization (WOCA) Conference, Lexington, KY, USA, May 4–7, 2009.

351. Truter W.F., Rethman N.F.G., Potgieter C.E., Kruger R.A., Re-vegetation of cover soils and coal discard material ameliorated with class F fly ash, Mining/ Reclamation II, World of Coal Ash Utilization (WOCA) Conference, Lexington, KY, USA, May 4–7, 2009.

352. Karmakar S., Mittra B.N., Ghosh B.C., Influence of Industrial solid wastes on soil-plant interactions in rice under acid lateritic soil, Agriculture I, World of Coal Ash Utilization (WOCA) Conference, Lexington, KY,USA, May 4–7, 2009.

353. Mittra B.N., Karmakar S., Swain D.K. and Ghose B.C., Fly ash a potential source of soil amendment and a component of infegrated plant nutrient supply system, Environmental I, International Ash Utilization Symposium, Center for Applied Energy Research, University of Kentucky, Paper-28, 2003.

354. Yunusa I.A.M., Eamus D., Silva D.L. D., Murry B.R., Burchett M.D., Skilbeck G.C. and Heidrich C., Prospect for coal ash in the management of Australian soils, Agriculture 2, World of Coal Ash Utilization (WOCA) Conference, Lexington, KY, USA, May 4–7, 2005.

355. Thenoux G., Acevedo L., Osorio A., Gonzalez M., Improved performance of soils stabilized with FBC ash, Aggregates/ Geotechnology III, World of Coal Ash Utilization (WOCA) Conference, Lexington, KY,USA, May 4–7, 2009.

356. Cohen H., Fly ash as a potential scrubber for low activity radioactive wastes, Environmental II, International Ash Utilization Symposium, Center for Applied Energy Research, University of Kentucky, Paper-51, 2003.

357. Mishulovich A., Evanko J.L., Ceramic tilesfrom high-carbon fly ash, Novel Applications 1, International Ash Utilization Symposium, Center for Applied Energy Research, University of Kentucky, Paper-18, 2003.

358. Kazouich G. and Kim A. G., The release of base metals during acidic leaching of fly ash, Environmental Aspects, International Ash Utilization Symposium, Center for Applied Energy Research, University of Kentucky, Paper-24, 1999.

359. Um N., Ahn J., Hang G., You K., Lee S., Cho H.C., Characteristics of magnetic substance classification from coal bottom ash using wet magnetic separator, Aggregates/ Geotechnology III, World of Coal Ash Utilization (WOCA) Conference, Lexington, KY,USA, May 4–7, 2009.

360. Si P., Qiao X.C, Luo Y., Song X.F., Yu J.G., Extraction of aluminum from combustion ash of coal spoil, New Product I, World of Coal Ash Utilization (WOCA) Conference, Lexington, KY,USA, May 4–7, 2009.

361. Zhou H M., Luo Y., Yu J.G., Qiao X.C,Feasibility of recovery Alumina from coal fly ash, Posters, World of Coal Ash Utilization (WOCA) Conference, Lexington, KY,USA, May 4–7, 2009.

362. Tranquilla J.M., MacLean J.H., Microwave carbon burnout(MCB): gas by products and development of specific metallic elements, Beneficiation, International Ash Utilization Symposium, Center for Applied Energy Research, University of Kentucky, Paper-107, 2001.

363. Fatima, Arroyo, Font O., Fernandz-Perira C., Querol X., Chimenos J.M., Zeegers H., Germanium and Gallium extraction from gasification fly ash : optimization for the up-scaling of a recovery process, Posters, World of Coal Ash Utilization (WOCA) Conference, Lexington, KY,USA, May 4–7, 2009.

364. Vassilev S.V., Menendez R., Diego, Alvarez and Borrego A.G., Multicomponent utilization of fly ash : dream or reality, Beneficiation 1, International Ash Utilization Symposium, Center for Applied Energy Research, University of Kentucky, Paper-12, 2001.

365. Michel L.J., Breighner E.B., Reclaiming western Maryland abandoned mines using coal combustion by-products, Posters, World of Coal Ash Utilization (WOCA) Conference, Lexington, KY,USA, May 4–7, 2009.

366. Koukouzas N.K., Zeng R., Perdikatsis V., Xu W., Emmanuel K. Kakaras, Mineralogy and geochemistry of Greek and Chinese coal fly ash : research for potential applications, Chemistry 5, World of Coal Ash Utilization (WOCA) Conference, Lexington, KY,USA, May 4–7, 2005.

367. Takao T., Kenji N., Masateru N., Jinmei L., Tatsuhiko S., Leaching test of coal fly ash for the landfill, Ash facility management I, World of Coal Ash Utilization (WOCA) Conference, Lexington, KY,USA, May 4–7, 2007.

368. Escobar Z., Lugo Y. and Hwang S., Biochemical response of landfill with manufactured aggregates as a daily cover, Aggregates/ Geotechnology III, World of Coal Ash Utilization (WOCA) Conference, Lexington, KY,USA, May 4–7, 2009.

369. Ward C.R., French D., Stephenson L., Riley K. and Li Z., Evaluating the interaction of coal ash leachates with rock materials for mine bachfill studies. Chemistry/ Minerology II, World of Coal Ash Utilization (WOCA) Conference, Lexington, KY,USA, May 4–7, 2009.

370. Yeheyis M.B., Shang J.Q. and Yanful E.K., Chemical and mineralogical transformations of coal fly ash after landfiling, Chemistry/Minerology III, World of Coal Ash Utilization (WOCA) Conference, Lexington, KY,USA, May 4–7, 2009.

371. Skodras G., Karangelos D., Anagnostakis M., Hinis E., Grammelis P., Kakaras E., Coal fly ash utilization in Greece, Posters, World of Coal Ash Utilization (WOCA) Conference, Lexington, KY,USA, May 4–7, 2005.

372. Maroto-Valer M. M., Taulbee D.N., Schobert H.H., Hower J.C. and Andresen J.M., Use of unburned carbon in fly ash as precursor for the development of activated carbons, New Products, International Ash Utilization Symposium, Center for Applied Energy Research, University of Kentucky, Paper-19,1999.

373. Behel D., TVA research on coal combustion by-product: use and environmental impacts, Environmental Benefits III, International Ash Utilization Symposium, Center for Applied Energy Research, University of Kentucky, Paper- 96, 2001.

374. Kochert S., Ricci D., Sorrenti R.and Bertolino M., Transforming bottom ash into fly ash in coal fired power stations, Cement & Mineralogy, World of Coal Ash Utilization (WOCA) Conference, Lexington, KY,USA, May 4–7, 2009.

375. Bitter J.D., Gasiorowski S.A.and Hrach F.J., Fly ash carbon separation and ammonia removal at tampa electric big bendBeneficiation/ Handling I, World of Coal Ash Utilization (WOCA) Conference, Lexington, KY,USA, May 4–7, 2009.

376. Petrik L.F.,White R.A., Klink M.J., Somerset V.S., Burgers C.L. and Fey M.V., Utilization of south Africa fly ash to treat acid coal mine drainage and production of high quality zeolites from the residual solids, Environmental 3, International Ash Utilization Symposium, Center for Applied Energy Research, University of Kentucky, Paper- 61,2003.

377. Stencel J.M., Ochsenbein M.P. and Cangialosi F., Automated foam index testing: A Quantitative approach to measure the capacity and dynamics during air entraining agent uptake, Cement or Concrete I, World of Coal Ash Utilization (WOCA) Conference, Lexington, KY,USA, May 4–7, 2009.

378. Sear L.K.A., Weatherley A.J., Dawson A., The environmental impacts of using fly ash – the UK products'perspective, Environment 2, International Ash Utilization

Symposium, Center for Applied Energy Research, University of Kentucky, Paper-20, 2003.

379. Yunusa I.A.M., Eamus D., Silva D.L.D, Murray B.R., Burchatt M.D., Skitberk G.C and Heidrich C., Prospects for coal ash in the management of Australian soils, Agriculture 2, World of Coal Ash Utilization (WOCA) Conference, Lexington, KY, USA, May 4–7, 2003.

380. Fernandez-Pereira C., Luis F. Vilches Arenas V., Miguel J.V.P., Rodriguer-Pinero and ValleJ.O.D. , Production of plates based on coal fly ash for their use as insulating materials in doors and fire break walls, New Products I, International Ash Utilization Symposium, Center for Applied Energy Research, University of Kentucky, Paper- 55,2001.

381. Lederman H.C.E., Pelly M.W.I., Polat M., Fly ash as a potential scrubber for low activity radioactive wastes, Environmentl II, World of Coal Ash Utilization (WOCA) Conference, Lexington, KY, USA, May 4–7, 2003.

382. Hemmings R.T., Cornelius B.J., Yuran P.and Wu M., Comprarative study of lightweight aggregates Aggregates/Geotechnology V, World of Coal Ash Utilization (WOCA) Conference, Lexington, KY,USA, May 4–7, 2009.

383. Diaz E.I., Allouche E.N. and Eklund S., Synthesis of zeolites from fly ash in a pilot plant scale: examples of potential environmental applications, Chemistry and Mineralogy, World of Coal Ash Utilization (WOCA) Conference, Lexington, KY,USA, May 4–7, 2009.

384. Dockter B.A., Eylands K.E. and Hamre L.L., Ceramic tiles from high-carbon fly ash, Novel Applications 1, International Ash Utilization Symposium, Center for Applied Energy Research, University of Kentucky, Paper- 56,1999.

385. ArenasL.F.V., Constantino Fernandez-Pereira C. L. F., Valle J.O.D. and Parapar J., Use of fly ash in a sprayed mortar for the passive protection against fire of metallic structures, Posters, International Ash Utilization Symposium, Center for Applied Energy Research, University of Kentucky, Paper- 56, 2001.

386. Karayigit A.I.and Gayer R.A., Characterisation of fly ash from the Kangal power plant, Eastern Turkey, Chemistry and Mineralogy I, International Ash Utilization Symposium, Center for Applied Energy Research, University of Kentucky, Paper- 4, 2001.

387. Fernandez I.D., Fernet L., Khahl C.A., Endres J.C.T., Maegawa A., Crystaline microstructure modification of Brazilian coal ash with alkaline solution, Chemistry and Mineralogy I, International Ash Utilization Symposium, Center for Applied Energy Research, University of Kentucky, Paper-88, 1999.

388. Kochert S., Ricci D., Sornenti R. and Bertoline M., Dry botton ash handaling increases marketability, LOI/Beneficiation/Handling I, World of Coal Ash Utilization (WOCA) Conference, Lexington, KY,USA, May 4–7, 2009.

389. Cedzynska K., Izydorezyk Z.K.M. and Sroczynski W., Plasma vitrification of waste incinerator ashes, Posters, International Ash Utilization Symposium, Center for Applied Energy Research, University of Kentucky, Paper-11, 1999.

390. Moutsatsou A., Itskos G.S., Koulouzas N. and Vounatsos P.P., Synthesis of aluminum based metal matrix composites (MMCS) with lignite fly ash as reinforcement material, New Product I, World of Coal Ash Utilization (WOCA) Conference, Lexington, KY,USA, May 4–7, 2009.

391. K. Koukouzas N.K., Zeng R., Perdikatsis V., Xu W., Kakaras E.K., Mineralogy and Geochemistry of Greek and Chinese coal fly ash : research for potential applications, Chemistry 5, World of Coal Ash Utilization (WOCA) Conference, Lexington, KY, USA, May 4–7, 2005.

392. Richard A Kruger and Japie E. Krueger, Historical development of coal ash utilization in South Africa, Policy 3, World of Coal Ash Utilization (WOCA) Conference, Lexington, KY, USA, May 4–7, 2005.

393. Kreger R.A., Hovy M.and Wardle D., The use of fly ash fillers in rubber, New Products, International Ash Utilization Symposium, Center for Applied Energy Research, University of Kentucky, Paper-72, 1999.

394. Lederman H.C.E., Werner M., Pelly H., Polat M., Synergetic effect of coal flyash as a scrubber to acidic wastes of the phosphate fertilizers industry, Novel Application 3, International Ash Utilization Symposium, Center for Applied Energy Research, University of Kentucky, Paper-50, 2003.

395. Cohen H., Seger G., Liberman R.N., Fly ash as a potential scrubber for low activity radioactive wastes, Environmental II, World of Coal Ash Utilization (WOCA) Conference, Lexington, KY,USA, May 4–7, 2009.

396. Tanosaki T., Watanabe Y., Ishkawa Y., Nambu M., Lin J., Yu Q., Nagataki S., Characterization of east Asian fly ash by polarization microscope, Posters, World of Coal Ash Utilization (WOCA) Conference, Lexington, KY, USA, May 4–7, 2009.

397. Bumrongjaroen W., Isabelle S., Davis J.M. and Richard A.,Livingston R.A., Characterization of Glassy phases in fly ash by particle SEM, Chemistry/Mineralogy II, World of Coal Ash Utilization (WOCA) Conference, Lexington, KY, USA, May 4–7, 2009.

398. Bumrongjaroen W., Muller I.S., Schweitzer J., Livingston R.A., Application of glass corrosion testes to the reactivity of fly ash, Chemistry/ Mineralogy II, World of Coal Ash Utilization (WOCA) Conference, Lexington, KY, USA, May 4–7, 2007.

399. Putilov V.Y., Putilova I.V. and Lunkov A.M., Some aspects of implementing ecologically sound ash removeal technologies at reconstruction of coal-fired power plants in Russia, Ash Facility Management II, Chemistry/Mineralogy II, World of Coal Ash Utilization (WOCA) Conference, Lexington, KY, USA, May 4–7, 2007.

400. Mahmud M.N., Marotovaler M. M. and Brandwood R., Characterization of pulverized fuel ash derived from coal and biomass cofiring and their utilization in to value added products, Benefiation/ Handling I, Chemistry/Mineralogy II, World of Coal Ash Utilization (WOCA) Conference, Lexington, KY, USA, May 4–7, 2007.

401. Querol X., Umana J.C., Felicia plana, Alastuey A., Lopez-Soler A., Medinaceli A., Valer A., Domingo M.J., Garcia-Rojo E., Synthesis of zeolites from fly ash in a pilot plant scale : example of potential environmental applications, Chemistry/ Mineralogy II, International Ash Utilization Symposium, Center for Applied Energy Research, University of Kentucky, Paper-12, 1999.

402. Meij R.and Berg J.V., Coal fly ash management in Europe: trends, regulations and health & safety aspects, Keynote-Plenary, International Ash Utilization Symposium, Center for Applied Energy Research, University of Kentucky, Paper- 108, 2001.

403. Ciccu R. , Ghiani M., Muntoni A., Serci A., Peretti R., Zucca A., Orsenigo L.G., Quattroni G., The Italian approach to the problem of fly ash, Processing, International

Ash Utilization Symposium, Center for Applied Energy Research, University of Kentucky, Paper-84, 1999.

404. Gitarani M.W, Fatoba O. O., Nyamihingura A., Petrik L.F, Vadapalli V. R. K., Nel J., October A., Dlamini L., Gericke G., Mahlaba J. S, Chemical weathering in a dry ash dump: an insight from physicochemical and mineralogical analysis of drilled cores, Environmental IV, World of Coal Ash Utilization (WOCA) Conference, Lexington, KY, USA, May 4–7, 2009.

405. Mishra A., Kumar S., Patel S., Gustin F., Resilient moduli and structural layer coefficient of fly ash stabilized recycled asphalt base, Aggregates/Geotechnology I, World of Coal Ash Utilization (WOCA) Conference, Lexington, KY, USA, May 4–7, 2007.

406. Buckley T.D., Debra F. Hassett P., Sager J., Ward J., Review of state regulations, standards and practices, related to the use of coal combustion product: taxes review case study, Policy 3, World of Coal Ash Utilization (WOCA) Conference, Lexington, KY, USA, May 4–7, 2005.

407. Wieckowska J. and Pietraszkiewicz W., Evaluation of processed bottom ash for use as lightweight aggregate in the production of concrete masonry units, Aggregates I, International Ash Utilization Symposium, Center for Applied Energy Research, University of Kentucky, Paper-43, 2005.

408. Drozhzhin V.S., Danilin L.D., Ya M., Shpirt, Kuvayev M.D., Potemkin G.A., Pikulin I.V., Formation processes of hollow microspheres in the fly ash from electric power stations, Posters, World of Coal Ash Utilization (WOCA) Conference, Lexington, KY, USA, May 4–7, 2005.

409. Putilov V. Ya., Putilova I. V., Modern approach to the problem of utilization of fly ash and bottom ash from power plants in Russia, Policy 2, World of Coal Ash Utilization (WOCA) Conference, Lexington, KY, USA, May 4–7, 2005.

410. Reynolds K.A. and Surridge A.K.J., Ash microbiology: A molecular study, Mining/Reclamation I, World of Coal Ash Utilization (WOCA) Conference, Lexington, KY, USA, May 4–7, 2009.

411. Kochett S., Ricci D., Sorrenti R. and Marco Bertolino M., The manufacture and evaluation of an artificial soil (SLASH) prepared from fly ash and sewage sludge, Forest/ Soil Reclamation, World of Coal Ash Utilization (WOCA) Conference, Lexington, KY, USA, May 4–7, 1999.

412. Zhai G., Yang Y., Gai G., Wu P., Scheetz B.E.and Roy D.M., Research on the whiteness and properties of moldified coal fly ash, New Product I, World of Coal Ash Utilization (WOCA) Conference, Lexington, KY,USA, May 4–7, 2009.

413. Maroto-Valer M. M., Taulbee D.N., Schobert H.H., Hower J.C. and Andresen J. M., Use of unburned carbon in fly ash as precursor for the development of activated carbons, New Products, International Ash Utilization Symposium, Center for Applied Energy Research, University of Kentucky, Paper- 19,1999.

414. Lan W., Yuansheng C., The application and development of fly ash in China, Cement and concrete III, World of Coal Ash Utilization (WOCA) Conference, Lexington, KY, USA, May 4–7, 2007.

415. Mazumder B., A process for production of silicon-carbide from spent pot liners of aluminum smelter plant, European patent No. 06019439.6–1218, 10th Nov. 2006.

Index